OXFORD MEDICAL PUBLICATIONS

Genes and the Mind

Genes and the mind

INHERITANCE OF MENTAL ILLNESS

BY

MING T. TSUANG, M.D., Ph.D

AND

RANDALL VANDERMEY, M.A., M.F.A.

OXFORD

OXFORD UNIVERSITY PRESS

NEW YORK   TORONTO

1980

Oxford University Press, Walton Street, Oxford OX2 6DP

OXFORD LONDON GLASGOW
NEW YORK TORONTO MELBOURNE WELLINGTON
KUALA LUMPUR SINGAPORE HONG KONG TOKYO
DELHI BOMBAY CALCUTTA MADRAS KARACHI
NAIROBI DAR ES SALAAM CAPE TOWN

British Library Cataloguing in Publication Data

Tsuang, Ming T
  Genes and the mind. – (Oxford medical
  publications).
  1. Mental illness – Genetic aspects
  I. Title   II. Vandermey, Randall   III. Series
  616.8′9′042   RC455.4.G4   80-40339

ISBN 0-19-261268-9

Typeset by Oxprint Ltd., Oxford
Printed in Great Britain
by R. Clay & Co. Ltd., Bungay, Suffolk

Preface

The governing purpose of this book is to help erase the stigma attached to the mentally ill and their relatives by giving the general public access to scientific evidence of the biological and genetic roots of many serious mental disorders.

The book is written in a style which makes it accessible to lay readers, yet its contents will be valuable to physicians, psychiatrists, medical students, residents, nurses, social workers, counsellors, and other health care professionals who may deal with the mentally ill. It is brief, and simply organized, not overly technical or clinical in form, and consistently humane in emphasis. It may be recommended by physicians or counsellors as a sort of frontline genetic counselling.

The book contains seven chapters, the first to the third being introductory, and the fourth to the seventh devoted to discussions of specific mental illnesses. Chapter One sets psychiatric genetic research in the perspective of the recent surge of interest in biological, genetic, and pharmacological approaches to mental illness. Chapter Two is a discussion of basic principles of human genetics and basic methods of psychiatric genetic research, specifically aimed at non-specialists. The chapter covers genes, chromosomes, mitosis, meiosis, patterns of inheritance, and methods of genetic research in psychiatry, especially family, twin, and adoption studies. Chapter Three describes the ramifications of genetic counselling in psychiatry, stressing who should counsel and be counselled, what happens in actual counselling sessions, stage by stage, the limitations of psychiatric genetic counselling, and how clients should interpret the concepts of risk, benefit, and burden.

Chapters Four to Seven discuss, in turn, four major psychiatric disorders – schizophrenia, manic depression, presenile dementia and Huntington's disease, and alcoholism, each of which is known to be, at least in part, genetically transmitted. These are the most common, some of the most severe, and the most intensively-studied psychiatric syndromes. Each chapter contains fictionalized case histories to dramatize the clinical features of the disorder in question. Each chapter offers a carefully and comprehensively re-searched survey of the most pertinent and most recent research establishing the presence of genetic factors in that given illness; briefly discusses theories concerning possible modes of genetic trans-mission for the given disorder; offers discussions, accompanied by tables, concerning the known life-time risks of the given illness from the immediate to the more distant relatives of an affected person, as established by empirical research; and ends with a concise summary of that chapter's contents.

We have selected schizophrenia, manic depression, presenile dementia, and alcoholism for discussion in the main chapters of this book because they are the most debilitating as well as the most intensively-studied psychiatric disorders. There is now a convincing body of evidence indicating the presence of genetic components in each of the four. These choices are not meant to imply that all other psychiatric disorders have strictly non-genetic origins. Evidence has begun to emerge showing that some of the other psychiatric dis-orders—neuroses, longstanding personality disorders, and social-opathy—may have genetic components within them as well. Furthermore, many forms of mental retardation show clear patterns of genetic transmission and are routinely the matter for genetic counselling. We would like to have made room in these pages for discussions of such disorders. But, in the case of neuroses, personality disorders and socialopathy, the evidence for genetic influence at present is still incomplete, and the diagnostic classifi-cations are so uncertain that any assertions that could be made would be far outweighed by hesitations and qualifications. In the case of mental retardation, which may take hundreds of forms, the genetic research is often conclusive and the body of research con-siderable—so considerable, in fact, that no respectable account of it could have been made within the confines of this book. Both areas of psychiatric genetic research deserve full treatment in another volume, written, as is this one, in language that the layman can understand.

In writing this book, we have incurred an unacknowledgeable debt to hundreds of researchers in psychiatry and genetics. Unable to mention them all, we have cited in the list of references at the end

Preface

those books and articles which have most directly contributed to our work. Some helpful suggestions for further reading are also given. We owe a more immediate debt to Ms. Jeannean Field for her secretarial skill without which this book could never have been completed. We also join gratefully the large company of those who have had special aid from, and been encouraged by, the positive vision of Ms. Marjorie Guthrie, President Emeritus of the Committee to Combat Huntington's Disease.

We are very grateful to the following authors for permission to reproduce illustrations: Dr. J. A. Fraser Roberts and Dr. Marcus E. Pembrey and Oxford University Press for Figs. 1, 2, and 3 from *An Introduction to Medical Genetics*; Dr. C. O. Carter and Penguin Books for Figs. 4 and 5 from *Human Heredity*; Mr. Sid Bernstein of Rockland Research Center, Orangeburg, New York for Plate 1; National Institutes of Health, Bethesda, Maryland for Plate 2 from The Overview (Vol. I) of the *Report to the Commission for the Control of Huntington's Disease and its consequences*. We are also much indebted to the staff of Oxford Medical Publications for their invaluable help in each step of the publication of this book.

The final manuscript of this book was completed while one of the authors (M.T.T.) was a recipient of the University of Iowa Developmental Assignment Award and Josiah Macy, Jr. Faculty Scholar Award for a sabbatical spent as visiting Professor at the Department of Psychiatry, University of Oxford, Warneford Hospital, Oxford, England, from September 1979 to August 1980, and the other author (R.V.M.) has been completing a four-year tenure as a Danforth National Teaching Fellow in English Literature at the University of Iowa, Iowa City, Iowa, USA.

Oxford   MING T. TSUANG
and   AND
Iowa City,   RANDALL VANDERMEY
January, 1980

This book owes its
existence to the teaching
and inspiration of
Dr. Eliot Slater C.B.E.

Contents

Contents

1
Are mental illnesses inherited?

Here is a letter from a man in New York City whose 18-year-old son has schizophrenia. After five years of treatment in mental hospitals, the boy has not improved. 'Can you help?' asks the anguished father. 'Please give me any gleaning of insight into new chemical weapons against schizophrenia, any current research holding out hope for my son and me at our crossroads.' The letter looks mimeographed. A copy has probably been sent to many doctors, like a message in a bottle.

Another letter comes from a woman in San Francisco. Her mother is severely depressed, as are two of her sisters. She wants to learn whatever she can about the hereditary side of mental illness: 'I can't find any books on the subject and I fear for myself.'

And a distraught Chicago matron writes that her stepson, aged 27, is crippled by fear of getting Huntington's disease, an illness which killed his natural mother and his uncle. The stepmother inquires: 'What are the chances that he too will become ill?'

These letters ask about different mental illnesses, but a single, implied question runs through each one: Is it true that mental illnesses have a biological basis?

This book answers YES, on the evidence of hundreds of studies substantiating the claim that genetic factors are at work in many major psychiatric syndromes.

The biological evidence

The top research journals in psychiatry today are full of articles exploring biological mechanisms which appear to be contributing influences in schizophrenia, manic depression, alcoholism, dementia, neuroses, antisocial personality, and many other mental disorders. These articles discuss molecular, functional, and structural properties of the brain and how they are implicated in various mental disorders. They reveal something about the role of neurohormones in disordered mental activity, about changes in enzyme

1

activity and cell metabolism. They help to explain changes in blood and urine associated with some mental illnesses. They relate mental illnesses to changes in skin response, eye movement, and brain wave activity. They explore complex relations between some mental illnesses and the body's immune response systems, the mechanisms which decide whether your body will accept or reject a graft of foreign tissue. And these studies bring us closer, bit by bit, to an understanding of the genes and chromosomes as they contribute to the development and hereditary transmission of mental disorders.

This mountain of contemporary research is evidence of a return swing of the pendulum in psychiatry. The 1940s and 1950s ushered in a preoccupation exclusively with psychological, environmental, and behavioural aspects of mental illness. But new biological discoveries and technology since the 1950s have returned psychiatry to the interest in biological and genetic aspects of mental illness which characterized it shortly after the turn of the century. The challenge to psychiatry today is to push the new biological discoveries through to their conclusions while opening up investigation of the complex interactions between psychosocial and biological influences in mental illness.

These are major developments in psychiatry, and anyone who has a stake in mental health—mentally ill patients, their families and friends, pastors, social workers, psychological, occupational, or marriage counsellors, medical students and residents, nurses, general practitioners, and psychiatrists—should know about them.

Biological research in psychiatry

One benefit of the advancement in biological research is that at last psychiatry has reached a point where discovery of the causes of some mental disorders seems to be within reach—a glimmer of light can be seen at the end of the tunnel. Biological research in psychiatry is still gaining momentum after two decades of cautious progress. But if research continues to accelerate at its present rate, especially in the areas of schizophrenia and manic depression, which are now under the heaviest assault, major biochemical and genetic breakthroughs can be hoped for in the lifetimes of many people now living.

Another benefit (and, in part, a cause) of the shift towards biological research is the discovery of psychotropic drugs like lithium carbonate for the treatment of mania, tricyclic antidepressants for depression, and major tranquillizers (phenothiazines, butyrophenones, etc.) for schizophrenia, to name a few examples.

Back to the community

These medications have changed mental health care in a way that strongly affects mental patients and the society around them. The new psychotropic drugs are capable of reducing symptoms of manic depression or schizophrenia in some cases so that thousands of mentally ill patients who might once have been subjected to long, costly, and often fruitless hospitalizations now can live at home receiving medication on an out-patient basis.

The back-to-the-community movement is reflected in the most recent statistics compiled by the National Institute of Mental Health (1977). These show that the number of first admissions to state and county medical hospitals in the United States was steadily on the increase after 1962, peaking in 1969 at 163 984, an increase of 34 286 in the space of seven years. But between 1969 and 1975, the number decreased by 43 284, bringing first admissions to their lowest level in 13 years, despite an overall rise in the US population during that period. In the last few years the rate of first admissions has not returned to its former height. The reintegration of the mentally ill into the family and community has in part been mandated by legislative action at the municipal, state, and national level, but the availability of effective biomedical treatments has done much to make the change possible.

Biological research into the fundamental causes of mental illnesses and research on the effects of psychotropic drugs reinforce one another. To learn why a certain drug works, researchers have to explore basic body chemistry. But discoveries in this area sometimes provide the key to new and more effective drug treatments, and so the cycle goes on.

From these and other angles of research psychiatry is gaining information which will enable it to distinguish more precisely between one mental illness and another. For example, since the discovery that manic symptoms responded favourably to lithium therapy, it has been found that there are some manic patients who do not respond to lithium. It has also been discovered that although most depressed patients do not improve on lithium, there are some for whom lithium works as an effective protection against the recurrence of symptoms. Those manic and depressed patients who respond unexpectedly favourably may represent a distinct subtype of manic depression, perhaps one which is genetically distinct from other subtypes. It is easily conceivable that subtypes of manic depression, schizophrenia, or other mental disorders would cut across diagnostic distinctions currently in use.

When it becomes possible to define pure subtypes of a mental illness, it will be easier for researchers to select pure subgroups for study than has so far been possible. With this advantage, researchers will be several steps closer to discovering the biological, including the genetic, foundations of these diseases.

Environment and heredity

The biological approach to mental illness does not oppose the idea that your parents, your home life, your diet, your friends, your own behaviour, and other environmental conditions may have a significant impact on the state of your mental health. That would be like denying that rain, sunshine, insects, and weeds have something to do with the way your garden grows. Although we have many psychological, behavioural, and other environmental theories to choose from, we have little knowledge as yet about which environmental factors might play important roles in various mental illnesses.

There are hundreds of paperbacks on the market proclaiming in bold letters how *you* can diet, exercise, or reason your way to better mental health, or how to rear your children so that they will be free of anything from tantrums to depression. But some of the most plausible-sounding of these books are in reality no more than a combination of guesswork, strongly held opinions, and hop-skip-and-jump research. They do not contain preventions or cures for the major psychoses and neuroses. Nor can theories of mental illness based only on concepts of environmental stresses and damaging patterns of rearing adequately explain the causes of the most serious mental disorders.

A purely environmental theory cannot for instance account for the results of adoption studies (summarized in Chapters 4, 5, and 7). Adoption studies investigate families in which a child has been removed from the parents at birth and raised with biologically unrelated adoptive parents. Since these studies eliminate almost all possibility of physical, psychological, or social exchange between the biological parent and the adopted-away child, we have to assume that if any more than a random degree of similarity appears between the two with respect to a certain trait, the increase must be caused by elements of the child's biological constitution inherited from the natural parent before birth.

In a major adoption study of schizophrenia (Kety *et al.* 1975), the close biological relatives of schizophrenic adopted children showed almost twice the rate of schizophrenia found in the biological relatives of normal adopted children, though all the children in both groups had alike been removed from their natural homes at or shortly after birth. Again, increased depression is found in adopted-away children of depressive parents (Cadoret 1978), and increased alcoholism is found in the adopted-away sons of chronic alcoholics (Goodwin *et al.* 1973). Results like these strongly suggest that there are important inherited biological factors in all these disorders.

Focus on psychiatric genetics

This short book is not intended to cover the whole gamut of research into biological factors in mental illnesses. Such a task would be gargantuan. The book would be as fat as the Manhattan and London telephone directories together and would include almost as many tongue-twisting names.

Our purpose, instead, is to focus on one pivotal area of biological research in psychiatry, namely, the study of genes and their role in the development and transmission of mental illnesses.

How do we know that some mental illnesses are genetically inherited? The evidence has come from several different types of study.

Family, twin, and adoption studies

The first clue is that many mental illnesses—schizophrenia, manic depression, alcoholism, neuroses, antisocial personality, and others—tend to concentrate according to type in families. Diseases which 'run in the family' behave the way one would expect genetic traits to behave. But some human characteristics that run in families—for instance, a love of music or tennis, or a tendency to be untidy—do not necessarily have a direct connection with the genes. A familial pattern of a given trait is thus no proof that the trait is genetically transmissible.

For firmer proof, reseachers have turned to one of the classic techniques in genetic research: twin studies. Identical-twin partners have identical set of genes, while fraternal (non-identical) twins have only half of their genes in common. Therefore, it is to be expected that when an identical twin receives a genetically inherited trait, the other partner should receive it too, all other things being equal. That is, the pair should be 'concordant' for the trait. Further, the rate of concordance for a given trait ought to be much higher for identical-twin pairs than for fraternal pairs.

This pattern has been found in twin studies not only of schizophrenia, but also of manic depression, alcoholism, sociopathy, and neuroses, indicating the presence of genetic components in each of these diseases.

But twins, like other children, are ordinarily reared in the same home environment with their parents and siblings. This means that twin concordance rates reflect both genetic and environmental factors stirred through each other. The twin study is not designed to separate the two. Adoption studies, on the other hand, are specifically designed to do so. They have provided some of the strongest available evidence of genetic factors operating in schizophrenia, manic depression, alcoholism, and antisocial personality.

Now that 'maps' are being made of the positions of genes on human chromosomes (though so far only a relative handful of genes have been precisely located), new possibilities are arising for studies called linkage studies, which can determine whether genes for a given disorder lie on a chromosome in close proximity to known genetic markers, such as the genes for colour blindness or certain blood types. If two genes lie close together, the two traits they cause are likely to be inherited together. Finding clear evidence of linkage between a given mental illness and a trait caused by a known genetic marker would provide hard evidence that the mental illness is genetically caused and would lay to rest any doubts left by loopholes in twin and adoption studies. However, these are techniques of the near future; linkage studies have been attempted, especially in the areas of manic depression and alcoholism, but no conclusive results have emerged from them yet.

The need for psychiatric genetic counselling

The hereditary aspect of mental illnesses used to be a thing that was whispered about or kept as a family secret, but now with the achievement of a new sophistication in biological and genetic studies and with the promise of major advances to come, information about the hereditary nature of specific mental disorders is being discussed— in doctors' surgeries, in local communities, and through the news media. The letters at the opening of this chapter represent some of the natural concerns of people who have picked up bits of this information.

Relieving anxieties

Are all mental illnesses hereditary? If I have mentally ill relatives, will I also be affected? Should I break off my engagement if a mental illness runs in my fiancé's family? If I have a mental illness, will my children inherit it too? Should we bear more children if one of us has developed a mental illness? Can a hereditary mental illness ever be cured or successfully treated? Is there anything I or my children can do to avoid inheriting a mental illness?

Questions like these can betoken a simple desire for information. But often, since they go to the heart of some basic human concerns, they are also fraught with anxiety. Such worries often afflict people in the prime of life, undermining their self-esteem, tarnishing their private dreams, and straining their primary human relationships. Many fundamental human satisfactions spring from the life and health of the family, but it is precisely these satisfactions which seem to be most threatened by prospects of the recurrence of a mental disorder.

If nothing at all were known about genetic components in mental illnesses, the worry alone would justify genetic counselling for persons affected with mental disorders and members of their families. Worry and the confusion that often accompanies it are problems in themselves in so far as they may become barriers to healthy self-acceptance, family harmony, and constructive decision-making.

But enough is already known about genetic components in many psychiatric syndromes to provide sufferers with understanding and substantial guidance. The initial stages of counselling—clarifying the problem, taking a thorough family medical and psychiatric history, firmly establishing the diagnosis of the illness in question—when conducted by an experienced counsellor, can in themselves help to bring order and direction out of family disarray.

Obstacles in psychiatric genetic counselling

Compared to genetic counselling for many non-psychiatric disorders, psychiatric genetic counselling is at a disadvantage. First, major illnesses like schizophrenia and manic depression often do not appear until after the victim has reached childbearing years, too late for genetic counselling to influence decisions related to family planning. Disorders with a typically late age of onset also add uncertainty to the process of drawing up an accurate family psychiatric history. Secondly, there are no known medical tests to detect potential victims of unaffected carriers of most mental disorders, as there are for many other diseases. Finally, the specific mechanism of genetic inheritance is unknown for practically every inheritable mental disorder. The only clear exception to this rule is Huntington's disease, which is known to be transmitted by a single dominant gene and follows classic rules of inheritance. The patterns of most other mental illnesses either fit no classic theory or fit several alternative theories of transmission. Without a clear understanding of the mode of transmission, it is impossible for the genetic counsellor to calculate exactly the likelihood that the relative of a mentally ill person will be affected with the disease.

The conditions are much more favourable for genetic counselling when a chromosomal disorder like Down's syndrome is concerned. While a baby is still in the mother's uterus, the presence of Down's syndrome can be detected. Amniotic fluid is withdrawn by a process called amniocentesis; cells floating in the fluid are then stained so that the chromosomal material can be analysed under the microscope. Not only Down's syndrome but all the other major chromosome disorders and many serious biochemical disorders as well as structural abnormalities of the brain and spinal column—about a hundred genetic diseases in all, many of them causing severe retardation or death—can be detected prenatally by this and other methods. Thus, instead of tenuous risk estimates, the parents can be offered a clear yes or no early in the pregnancy. Many other serious genetic diseases—cystic fibrosis, haemophilia, Tay-Sachs disease, and sickle-cell anaemia, to name a few—can be detected early in life and follow clearly determined hereditary patterns, making exact risk calculations possible.

In the absence of exact risk calculations

Despite the fact that exact risk calculations are unavailable for mental disorders, the hands of the psychiatric genetic counsellor are not tied. He or she can make fairly accurate estimates of risk compiled from family studies. In these studies researchers have investigated the relatives of persons with a given disorder, noting how often the same disorder occurs in each different class of relatives. In the more heavily studied syndromes such as schizophrenia and manic depression, the breakdown can become quite specific, differentiating between first-, second-, and third-degree relatives (i.e. immediate blood relatives, blood relatives one step removed, and those two steps removed), between males and females, and between different types of family members (parent, sibling, or child). The estimates, offered in the form of percentages, indicate the proportion of relatives in a given class who are affected by a given disorder.

Empirical risk estimates, as these are called, have nothing to tell us about the causes of the mental illnesses they measure. They reflect rates of recurrence heedless of whether the disorders stem from gene action, environmental conditions, or (as is very often the case) from complex interactions between the two. Nor do empirical risks take into account the possibility that some major psychiatric disorders may in fact be several distinct types of illness travelling under one name.

With such reservations in mind, however, it may be acknowledged that empirical risk estimates are the best figures available for most mental disorders, pending further breakthroughs in biochemical, genetic, or psychosocial research. They can serve a useful purpose in genetic counselling, as the following case illustrates.

A 35-year-old man consulting a genetic counsellor after his mother was hospitalized for depression persisted in believing that he had a 50–50 chance of becoming depressive himself. His simple (but error-ridden) logic was that since there were, to his mind, only two possible outcomes (either he would inherit the disorder or he would not), he could translate these two possibilities into a straight heads-or-tails probability. He was somewhat relieved when the counsellor informed him that the probability of his inheriting his mother's disorder, based on empirical studies, was far less than he had supposed.

But overestimation of risk is only one problem. In some cases it happens that one of the members of a family in counselling lightly dismisses all thoughts about risks of recurrence. Where such individuals may dismiss statements from the counsellor or relatives as ungrounded personal opinions or, worse, as attempts at manipulation, they will sometimes respect figures compiled from studies around the world.

9

On the other hand, people who are still taxed by the physical and emotional stresses of living with or fearing the recurrence of a mental illness in their family are likely to find lists of figures alone difficult to interpret.

It is up to the counsellor to assist his clients towards a clear understanding of the figures and to help them to take a positive course of action through the human conflicts arising from their situation. Sensitive, honest, and professional genetic counselling requires that the counselling transaction must be adjusted uniquely to each unique set of needs by a counsellor who tries to consider the total welfare not only of each individual in counselling but of the family unit and, if family planning is at issue, of the unborn child.

Fighting the stigma of mental illnesses

Do an experiment. Walk past a general hospital, then past a psychiatric hospital. Better still, if you can find the two facing each other on opposite sides of the road (as one can in Iowa City), walk along the road between them. How do you feel as you pass each of them?

Do you feel a certain warm glow as you pass the general hospital, warmth rising from a combination of positive associations with the place? A feeling of sympathy with the medically ill, hope for their recovery, trust in the heroism of their doctors? Perhaps it amounts to a sense that all of them are benefiting from the best of available technology and physical comforts.

And as you pass the psychiatric hospital? A vague sense of foreboding, perhaps, a slight but unaccountable hint of fear? Just what does go on in there, anyway? Perhaps one finds oneself watching even the squirrels on the lawn a little too carefully—as if for signs of odd behaviour? Or listening a little too hard, as if to catch discordant sounds from within? One's steps become more tightly measured as one suppresses the impulse to run.

Exaggerations? Perhaps, but with just enough truth in them to expose the real and hurtful stigma often attached to mental illnesses. The stigma is not only conveyed by prejudiced observers—a sense of it is absorbed by psychiatric patients and their families as well. One can watch it surface in casual social contacts between psychiatrists and their patients. A family doctor bumping into one of his patients in the supermarket will be treated to a recital of the patient's progress: 'I've been keeping up faithfully with the insulin, doctor!' or 'I'm feeling fine now, but I've had a touch of flu since my operation.' It is not unusual, on the other hand, for psychiatric patients meeting their psychiatrist in public to walk past without even a greeting.

This book is written in the belief that spreading information about mental illnesses—particularly about new advances made by research into their biochemical and genetic foundations—is the most effective way of combating that prejudice born of ignorance which continues to load an extra burden of suffering on the mentally ill and their relatives.

But let us test that claim for a moment. If genetic components have been discovered in major psychiatric disorders, why should the knowledge of them remove the stigma? Don't these discoveries rather confirm the worst that has been suspected? Don't they seal the fate of persons unfortunate enough to be afflicted?

To be sure, some people's fears must be confirmed by the knowledge that the mental disorder in their family is inheritable. But such knowledge, however saddening, can provide mental equilibrium and equip the knower for rational action. A stigma cannot do that; it thrives in an atmosphere of mystery, fear, suspicion, or guilt. A stigma brings about paralysis, not action. Although it may have started from a fragment of truth, it tends to snowball, snatching up false rumours, prejudices and other mental debris in the process. The stigma attached to mental illnesses (and sometimes ignorantly equating them with mental retardation and physical handicaps) treats all of them as if they were inexorably inheritable. But psychiatric genetic studies have shown that this is not so. They may confirm the worst fears of some people, but they will assure many others that their worst fears were unfounded.

A number of facts must become common knowledge before we can get rid of the stigma attached to mental illnesses.

First, *mental illness is not all one thing.* Just as there are many distinct physical diseases, so there are many distinct mental disorders, having different causes and requiring different treatments. Not only are the major syndromes distinct from one another, but in all likelihood, numerous different disorders are grouped within some of them sharing the same symptoms but having different causes.

Secondly, *not all mental illnesses are inheritable.* Many of the medical disorders which have been identified have at least two forms, one apparently inheritable and the other non-inheritable, a

11

relatively short-term reaction to environmental stresses or an anomalous condition of no known cause. For example, a psychosis traceable to abuse of amphetamines may be indistinguishable from schizophrenia except on the basis of family history, the victim's past experience with drugs, and the course of the illness. Many depressions which occur late in life show no evidence of being passed along within the victim's family. Many other examples could be chosen.

Thirdly, *inheritable diseases are not inherited 100 per cent of the time*. When a parent has an inheritable mental illness—let us say, for simplicity, that the disease follows classic rules like Huntington's disease—each of that parent's offspring has a chance of inheriting the disease and a chance of escaping it. The chance of inheriting Huntington's disease from an affected parent is relatively high—as high as 50 per cent for children of one affected parent. In almost all other inheritable mental illnesses, the child's chances of inheriting the disease are several notches lower, in the area of 10–15 per cent for children of one schizophrenic or manic-depressive parent, though somewhat higher if more than one person in the family is affected.

But even when the damaging gene or genes are inherited the likelihood of the person developing the damaging symptoms of the disease is not necessarily as high as 100 per cent. Some genes are dependent on age and may never act while their owner is living. Others may fail to penetrate the body's metabolic pathways and thus may leave the owner essentially normal. Some genes may be closely interrelated with factors in the owner's environment, failing to manifest themselves if insufficient environmental stimulus is present. This may be especially true of mental disorders produced by the joint action of many genes.

Finally, *inherited mental disorders are not all incurable*. Even though some psychiatric disorders, such as dementia, often follow a deteriorating course leading to death, other disorders respond to treatment. For example, phenylketonuria—a rare genetic disease appearing in the first year of life which may involve psychiatric complications—can usually be controlled by early supervision of the infant's diet. Symptoms of mania often moderate when the affected person is treated with lithium carbonate. Persons with inherited depressive disorders are often able to function normally when treated with antidepressant drugs.

In the past two decades the number of effective drug treatments

Fighting the stigma of mental illnesses

for mental disorders has greatly increased as a result of advancements in biological research in psychiatry. Today the hope is greater than ever before that those mental illnesses which have always been resistant to any kind of treatment may someday be controllable through new biomedical techniques.

The discovery of genetic factors underlying mental disorders, rather than closing the door to effective treatments, opens the door wider.

2
Human genetics
for beginners

This chapter spells out in simple terms the basic facts of human heredity. These facts will help you to understand the language of psychiatric genetic research, to reason more knowledgeably about issues related to genetic diseases and their transmission within families, and to erase the damaging stigma so commonly attached to mental illnesses.

Readers who have taken a college-level course in genetics may want to skip this discussion and proceed to the following chapters, although a refresher course in basic genetic theory would probably do no harm. To most people, genes, chromosomes and other complex mechanisms of the body's 'inner space' seem about as far removed as the quasars, nebulas, and black holes of outer space. But this book does not require that its readers have doctoral degrees in molecular biology. Those who are able to gain a clear, non-technical grasp of some simple genetic terms and principles are more likely to recognize a popular misconception when they see one, more likely to comprehend the problems facing the victims of hereditary mental disorders and their relatives.

A little knowledge may be a dangerous thing, but where inheritable mental diseases are concerned, complete ignorance may be devastating.

Genes

Let us begin with a single gene, the body's smallest chemical unit of hereditary information, inside one of the body's billions of tiny cells.

A gene is a highly complex molecule of a substance called *deoxyribonucleic acid*, usually abbreviated DNA. Since the Nobel-Prize-winning discovery of Drs James Watson, Francis Crick, and Maurice Wilkins in 1953, scientists have known that this molecule is formed something like a ladder, with anything from less than one hundred to more than a thousand steps, coiled into a spiral-staircase, or helical, shape.

Genes

The composition and order of the steps give DNA its marvellous ability to make an exact copy of itself during cell division and to control the manufacture of basic biochemicals when the cell is in its normal state. Each ladder-step is made of two simple bases weakly attached together. Each side of the ladder is made of simple phosphate and sugar molecules firmly bonded to one another to form a spine-like structure; each of the phosphates or sugars is bonded on the inside to one of the protein molecules that forms half of a step. Figures 1 and 2 show that two strands of the sugars and phosphates

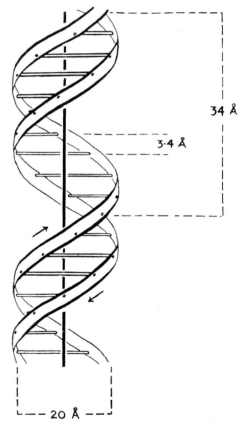

FIG. 1. Diagram illustrating the double helix of DNA. The dimensions are in Angström units (A). One Angström unit is 10^{-7} mm. The two ribbons symbolize the phosphate-sugar chains and the horizontal rods the bases holding the chains together. The arrows show that the sequence of bases goes one way in one chain and the opposite in the other. (Courtesy of Fraser Roberts and Pembrey.)

15

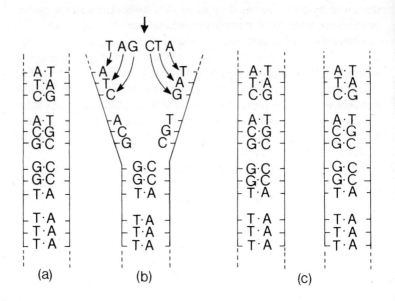

FIG. 2. Diagram showing how the double helix of DNA divides, giving two identical daughter helices (a) before division (b) during division (c) after division. (Courtesy of Fraser Roberts and Pembrey.)

are bonded by pairs of bases. Adenine is always attached to thymine, and cytosine always to guanine.

When the DNA replicates itself, it first splits the long way, much like a zip-fastener, each step breaking at the weak joint between the two bases. Now there are two separate strands, each bristling with protein half-steps which, chemically speaking, want to be completed with partners similar to the ones they had before. The cell then supplies appropriate new bases, phosphates, and sugars to the separate strands so that two identical new ladders are formed. Each of the two cells formed by cell division will receive one of these two newly manufactured DNA molecules from the original cell.

Each gene acts like a chemical mimeograph machine, using its coded sequence of phosphates and sugars to duplicate amino acid chains, which are called *polypeptides*. 'One gene, one polypeptide' is the general rule. This means that each gene has the specific function of producing only one polypeptide. However, a polypep-

16

tide may go out into the cell to combine or interact with polypeptides designed by other genes. Therefore, the secondary effects of one gene's action can be numerous and widely varied.

The actions of a gene will generally fall into one of two basic categories. Most genes are used to build proteins. They are called *structural genes* because they guide the manufacturing of materials that go into the structural organization of the cell. The second type are called *regulator genes*. Little is known about their specific functions in human beings, but it appears likely that these genes, rather than producing polypeptides for structural purposes, produce molecules that tell other genes when to turn on and off.

The genetic material of a cell is the basis for an amazingly complex self-regulating system which produces effects in the body, interprets feedback, reacts with new effects, and so on throughout the life cycle of the organism.

A gene almost always duplicates itself correctly, but on very rare occasions it will make a mistake—maybe a very small one, like missing one out of several hundred steps. After that, the DNA molecule will faithfully copy the mistake each time it reproduces itself, like a secretary using a flawed master stencil on her mimeograph machine. The error is called a *gene mutation*.

Because of a mutation, cells may start to produce a substance which does harm to the organism. Or the mutation may cause a change in the genetic code which the cell interprets as chemical gibberish. As a result, the cell may fail to make a certain necessary enzyme, thus harming the organism. Or, as in a hereditary disease like phenylketonuria, the cells may fail to break down given substances as they should, thus causing that substance to accumulate in the body to a harmful level.

Mutations are not always harmful. Some produce pleasing variations in human traits. Some even give the mutant an advantage in survival. But others may be harmful or lethal to the organism. Harmful genes, like normal ones, can be transmitted from parent to child at conception; they can survive in a population as fairly common alternatives to the normal gene.

The place of genes in the cell

To simplify this discussion of human heredity, we began by describing how a single gene is made and how its actions may affect the cell and the whole organism. This approach may have created the

17

impression that your whole genetic endowment is concentrated in one gene which is repeated in every cell, but that impression would be far from the truth. Each human cell (with a notable exceptions) does indeed contain a complete set of genetic material for the whole organism, but that complete set is composed of many, many genes.

Scientists still cannot accurately count the number of genes in each human cell. Genes are too small to be seen even under the most powerful electron microscope. Geneticists have already identified more than 1800 specific genes (McKusick 1977) and the number of known genes is growing every year. Large though this number may seem, however, it is only a tiny fraction of the probable total number of genes in a human cell. Most estimates fall in a range from 10 000 to 2.5 million!

How can so much genetic material fit within each single cell? The large numbers may be deceiving, for even in great quantities, genes use up very little space. Professor Curt Stern from the University of California at Berkeley, estimates that if we took the DNA from the egg and sperm cells which would be needed to produce each of the individuals in the next generation—let's say six billion people—and packed it all together, the whole bundle would have less volume than $1/10$ of an aspirin tablet!

Let us look at the structural details of the cell to understand where the genes are situated. Each normal body cell has an outside membrane called a *cell wall* containing a fluid called *cytoplasm.* Floating in the cytoplasm is a special compact structure called the *nucleus*, which looks like a miniature cell within the cell. The nucleus might be called the command headquarters, because it contains many structures, including the genes, which direct the manufacture and traffic of substances within the cell.

In the cell's normal condition, the genes are found within the nucleus on long, tiny, relaxed fibres called *chromatin.* But when a cell is preparing for its miraculous division into two new cells, the chromatin fibres contract into short, tightly coiled threads. When cells at this stage are fixed and stained in the laboratory, these threads take up stain very well and become clearly visible under the microscope. For this reason they have been given the name *chromosomes*, from the Greek words *chromos* and *soma*, meaning colour-body. Each chromosome may hold hundreds or thousands of genes.

Chromosomes

As recently as 1956, it was thought that human beings, like gorillas and chimpanzees, had 48 chromosomes in each cell. However, with improvements in staining techniques, the number of chromosomes in a normal human cell has been set beyond a doubt at 46.

It is possible, by microscopic photography, to take a picture of all the chromosomes in the nucleus just before cell division, when they are most visible. The images of all the chromosomes can then be cut out and arranged according to relative size and shape (see Figure 3). From this arrangement, called a *karyotype*, it is easy to see that not all chromosomes are alike. The smallest ones are only about one-fifth the length of the largest. Each of the chromosomes has two discernible segments or arms joined at a small, darker-staining point called a *centromere* (also called *kinetochore*). Some of the chromosomes, both short and long ones, have arms of equal length on each side of the centromere. Others have a short arm on one side of the centromere and a long arm on the other. Other chromosomes

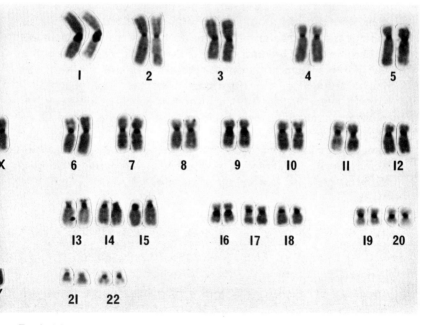

Fig. 3. A karyotype of a normal male showing the 22 pairs of chromosomes and the X and Y chromosomes. (Courtesy of Fraser Roberts and Pembrey.)

appear to have centromere located at the end, with only the tiniest arm visible on one side.

The autosomes

The karyotype reveals another important feature of the chromosomes: each one has a twin partner. At the time of conception, the mother and father each contributed one complete set of 23 chromosomes to the first whole cell, called the *zygote*. Thus, the zygote contained 23 matched pairs of chromosomes. All the cells descended from the zygote likewise received two of each chromosome, one descended from the mother and one from the father. Two chromosomes which form a matched pair are called *homologous chromosomes*. In each cell nucleus there are 22 pairs of homologous chromosomes referred to as *autosomes*. Twenty-two pairs, you notice, adds up to 44 chromosomes, but we said previously that each cell contains 46. The two remaining chromosomes, called *sex chromosomes*, are homologous in a female but not in a male. We will discuss their importance shortly.

Let us first look more closely at one pair of homologous chromosomes to consider how pairing influences the development of genetic traits in the individual. One member of the pair—let us call it chromosome Q—carries, say, a thousand genes, each with a specific biochemical function. These genes are not scattered haphazardly along the chromosome. In fact, each gene has a specific place, called its *locus*, where it appears on every chromosome of that type. If we would travel to another cell and look up that cell's copy of chromosome Q, we would find the same genes at the same loci. Every chromosome in the same nucleus as chromosome Q, with the exception of Q's homologous partner, will have different gene loci bearing genes which influence a different set of traits.

Now if we compare one locus on chromosome Q with the corresponding locus on Q's homologous partner, which we will call Q' (Q-prime), we might discover one of two things. Q' will either have the exact same gene at that locus, or it will have a slightly different gene coding for a variation on the same trait. For instance, the gene on Q might code for brown hair, while the gene at the same locus on Q' codes for the trait 'not-brown' (that is to say, blond) hair. When homologous chromosomes bear identical genes at a certain locus, the person is said to be *homozygous* for that genetic

20

trait; when they bear non-identical genes at that locus, the person is *heterozygous*. The various genes that may alternate at one locus are called *alleles* with respect to one another. Some genes have only one allele, while others have several. It is important to remember that some alleles may be harmful to their owner's health. For example, where one person has a gene which influences normal development of the central nervous system, another person may have the allele which causes the fatal central nervous system disorder called Huntington's disease. The fact that human genes may have *multiple alleles* helps to explain the great diversity of human traits.

But not all alleles are equally assertive. In the heterozygote, one of the alleles will overrule the other. The one that overrules is called a *dominant gene*, while the one overruled is called a *recessive gene*. A recessive gene will occasionally manifest itself as an observable trait, but that typically happens only when the homologous chromosome also has a recessive gene at that locus so that no dominant gene can move in and take over. Another way of putting it is that the dominant gene is one which will be expressed in both the homozygotic and the heterozygotic person, while the recessive gene will only be expressed in a homozygote. Dominance and recessiveness of genes further helps to account for the great diversity of human traits.

The sex chromosomes

We have been talking about the typical characteristics of the 22 pairs of autosomes and their cargo of genes. The remaining pair, called the sex chromosome, have been set aside because they play a unique role in the genetic life of the individual.

Sex chromosomes receive their name not because they have any particular sex appeal or because only wild Hollywood starlets have them, but because, among other duties, they determine each person's gender, male or female. There are two types of sex chromosome, a relatively large one labelled X and a very small one labelled Y which seems to bear only genes for male traits. If you are a normal woman, each of your cells contains two X chromosomes and no Y. If you are a normal man, each of your cells contains exactly one X and Y. Since the mother has only X chromosomes to bequeath to her offspring at their conception, the sex of the child depends on whether the father transmits an X or a Y chromosome (an event which is determined by chance). In this sense, the father determines the sex of his offspring.

Very few traits are know to be transmitted via the Y chromosome. Certainly no genes essential to human life can be carried on it, since women thrive quite well without one. But the X chromosome is the known vehicle for many genetic traits, which are said to be *sex-linked* or *X-linked*. Many sex-linked genes have been specifically identified. Some of these, such as the sex-linked genes for certain blood types and for colour-blindness are called *markers*, because they can be used to mark the relative position of other genes on the X chromosome.

In a female, an X-linked recessive gene may be overruled by a dominant allele on the homologous X chromosome. But in a male, recessive genes on the X chromosome are normally expressed as traits, since the small Y chromosome usually carries no alleles to overrule them.

Cell division

Ordinary cells in the human body multiply by a process called *mitosis*. For the purposes of this chapter, it is not necessary to go into the fine details of this process. Essentially what happens is that the cell makes a copy of the classified information in its nucleus and splits into two identical new cells, giving one full set of information to each. The whole process usually takes no more than an hour.

We have already discussed how DNA replicates itself by 'unzipping' and recomposing its two sides out of chemicals available to it in the cell. This process is at work during cell division when the chromosomes in the nucleus make copies of themselves. If the cell is stained and fixed at one of the stages of mitosis before the division is complete, one can see through the microscope that each of the chromosomes has doubled itself and is lying closely parallel with its copy, still joined to it at the centromere. In the next phase of mitosis, the centromeres line up across the midline of the cell. Each centromere splits in two, one half taking one copy of the chromosome and the other half taking the other copy. One complete set of chromosomes is dragged by its centromeres toward the opposite side, thus forming two complete sets of chromosomes identical to the original. Separate membranes surround each set, thus forming what will be the nuclei of the two new cells. Meanwhile, an invisible belt seems to tighten itself around the midsection of the cell, gradually squeezing, separating the cytoplasm into equal portions for each of the two new cells. Finally, the cell wall pinches off in the middle and closes around each of the newborn cells. The process of mitosis is complete. Each new cell now possesses the full set of hereditary information for that individual human being.

Meiosis: shuffling the deck of genes

The process of mitosis just described insures that each new cell formed by cell division receives 23 chromosome pairs—a full set. Now we must discuss two important exceptions to this rule: the egg cells formed in the female ovaries and the sperm cells formed in the male testes.

During the process of human reproduction, an egg and a sperm unite to form a fertilized egg, or zygote, the first complete cell of a new human being. It will divide into thousands of billions of cells before it is fully grown into a mature adult. The zygote has a complete set of genes and chromosomes by virtue of the fact that both the egg and the sperm which formed it contained exactly half as many chromosomes as ordinary cells. Fused at the time of conception, they together contributed the full set of genetic material.

How did the egg and sperm develop only half the normal number of chromosomes? By a process of reduction division called *meiosis*. For the purpose of this discussion, a sketch of the basic stages of meiosis should be sufficient. We will use the development of a single sperm cell as an example, keeping in mind that an egg is produced in much the same way.

The pre-meiotic sperm cell has 23 pairs of chromosomes (46 chromosomes in all) like any other normal body cell. During the process of meiosis, this cell will divide once, producing two new cells; both of these will divide again, so that four cells will be produced in the end. Each of the four will become a mature sperm cell.

During the first division, each of the 46 chromosomes first makes a copy of its two strands, then lines up beside its homologous partner (which has also doubled). While the pairs are lined up side by side like this, some of the chromosome strands may lap over or twine around the corresponding strands on their partner chromosome; in the process, two strands may exchange segments, each segment having the same loci but possibly different alleles at these loci. This process is called *crossing-over*. It is as if two large families decided to exchange some of their children. If we remember that in each homologous pair of chromosomes one has been donated from the maternal side and the other from the paternal side; we see that crossing over results in a shuffling of maternal and paternal genes. This random shuffling may take place any time a sperm or an egg cell is produced.

The number of different combinations of alleles that could be produced in this way from the 23 chromosome pairs of one parent cell is astoundingly large. Stern (1973) conservatively estimates that if a man produces 1 trillion sperm cells in his lifetime, this number represents only about one sixty-million trillion trillionth of the total number of possible combinations of alleles in a cell containing 23 chromosome pairs! And you wondered why your brothers and sisters are different from you. . . .

We left our doubled chromosomes side by side with their partners, preparing to separate. After crossing over is completed, the nucleus begins to split in two; at the same time, one doubled chromosome from each pair migrates toward one half of the splitting nucleus while its doubled partner migrates toward the other. When the cell division is complete, each of the two new cells contains its own

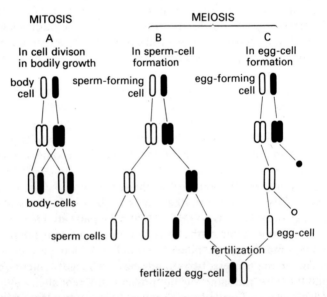

FIG. 4. Diagram showing the behaviour of a single chromosome pair in cell division. (Courtesy of Carter.)

Meiosis: shuffling the deck of genes

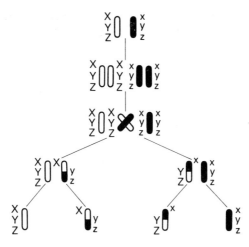

Fig. 5. Diagram showing how genes originally on the same chromosome may be separated by 'crossing-over' during meiosis. (Courtesy of Carter.)

nucleus, which in turn contains 23 doubled chromosomes. The homologous partners, after crossing over, have gone their separate ways.

Shortly after the first division, both newborn cells divide again. But this time the chromosomes do not duplicate before division—remember, they are already doubled. Instead, as the nuclei begin to divide, each doubled chromosome divides at the centromere, where the two copies had been joined like siamese twins. Thus, from each original centromere two new ones are formed, each joining the two arms of a single chromosome. One of each pair of centromeres then migrates toward one side of the dividing nucleus, dragging its chromosome with it, while the other centromere drags its chromosome in the opposite direction. When this division is completed, four cells stand where originally there was only one. Each of the four cells contains 23 single chromosomes (22 autosomes and one sex chromosome) or half the number of chromosomes in an ordinary cell.

The egg cell in the female is produced in very similar fashion, but the final result is not four eggs of equal size. Instead, one of the eggs retains a great deal of cytoplasm, making it one of the largest cells in the human body, while the other three are stunted cells with very little cytoplasm. These three stunted cells, called *polar bodies*, do not survive long.

The two most important facts to remember about meiosis are: (1) that meiosis reduces the number of chromosomes in the human germ cells (egg and sperm) to one-half the normal number so that when a sperm cell fertilizes an egg at conception, the fusion of sperm and egg nuclei will restore the complete set of human genes and chromosomes in the nucleus of the zygote; and (2) that during meiosis, crossing over and other features of chromosome behaviour are responsible for the random shuffling of maternal and paternal alleles, thus insuring that each offspring (except in the case of identical twins) will receive a different set of alleles from its parent.

Single-gene heredity

The classic models of human heredity are single-gene traits, i.e. traits communicated from parent to offspring by the transmission of dominant or recessive alleles at one locus on one of the chromosome pairs. Such traits usually follow predictable patterns of inheritance.

The first accurate description of single-gene heredity came from the work of a nineteenth-century Austrian monk named Gregor Mendel (1822–84), long before anyone knew about genes and chromosomes. Mendel bred and crossbred many generations of pea plants and observed how clearly-contrasted traits in the pea, such as tallness versus shortness or green versus yellow colour, were distributed among the offspring in succeeding generations. From these pioneering experiments he learned that some traits, which he called *dominant*, prevailed over others, while he called *recessive*. When a parent plant having a dominant trait was crossbred with a plant having the recessive trait, the first generation would inherit the dominant, not the recessive, trait. But when the dominant plants of the second generation were bred with each other, a variety of types were produced: some were recessive, some were dominant and able to bear only dominant offspring, and some were dominant but able to bear both types of offspring. This demonstrated that a dominant plant could be a *carrier* of a recessive hereditary factor without developing the recessive trait. Mendel painstakingly worked out the expected ratios of traits among the offspring of various combinations of homozygotic and heterozygotic parents. His summary analyses of these observations are now called 'Mendelian laws of inheritance'. They are still applicable to single-gene traits not only in pea plants but also in human beings.

Single-gene heredity

Like many other great advancements in science, Mendel's discoveries were ahead of his time. The value of his experiments went unacknowledged after his death until 1910. For most of this century, however, Mendel has been recognized as the father of the modern science of genetics. Simple single-gene dominant and recessive characteristics are still referred to as 'classic Mendelian traits',

Because chromosomes can be either autosomes or sex chromosomes and genes can be either dominant or recessive, four different patterns of transmission are possible in single-gene heredity: *autosomal dominant*, *autosomal recessive*, *sex-linked dominant*, and *sex-linked recessive*.

Autosomal dominant

An autosomal dominant trait will ordinarily be expressed if the dominant gene is present on either member of a chromosome pair. Some persons who possess a dominant trait are homozygous, having inherited the dominant gene from both their mother and their father. Such a person will only be able to transmit a dominant gene for that trait to his or her children. Therefore, every child will receive at least one dominant gene and will inherit the trait. If both parents are heterozygous, one out of four of their children will be homozygous for the dominant allele and one out of four will be homozygous for the recessive allele. Two out of four of their offspring will be heterozygous. But because a total of three out of four offspring will inherit at least one dominant gene, and because those who inherit a dominant gene normally express the trait, three out of four children of two heterozygous parents will inherit the dominant trait. For any child of an affected parent, the minimum likelihood of inheriting a dominant trait is one out of two. Autosomal dominant traits affect males just as often as females and typically do not skip generations before recurring in a family. Only one mental illness, namely Huntington's disease, clearly fits an autosomal dominant model of the classic type, where the dominant gene is expressed in 100 per cent of those who possess it.

Autosomal recessive

In autosomal recessive inheritance, the recessive characteristic appears only when the individual is homozygous for the recessive gene—in heterozygotes, the dominant allele on one of the two homologous chromosomes overrules the recessive allele's effects. The person who possesses both a dominant and a recessive allele at one locus without expressing the recessive characteristic is called a *carrier*. If both parents are normal but one is a carrier of a recessive gene, two out of four of their children will also be carriers, although none will develop the recessive trait. It is often estimated that every person is a carrier of at least four to eight recessive genes for serious genetic diseases. But since there are many possible recessive diseases, the likelihood that two people carrying the same recessive genes will meet, marry, and each transmit the recessive gene to a child (who would thus be homozygotic recessive and affected with the disease) is too small to be much cause for worry. If both parents are normal carriers, one out of four of their offspring will inherit a recessive gene from each parent and be affected. Two out of four will be normal carriers and one out of four will be a normal non-carrier.

In the event that a person with a recessive characteristic (who must therefore be homozygotic) marries an unaffected carrier of the recessive gene (i.e. a heterozygote), all of the children will receive a recessive gene from the affected parent, while two out of four will also receive the recessive gene from the carrier parent; thus, two out of four will be homozygotic and will inherit the recessive characteristic, while the other two out of four will be carriers and unaffected. In the less likely event that two persons, both homozygotic for a recessive gene, marry and raise a family, all of the children will be homozygotic and will inherit the recessive characteristic. Autosomal recessive characteristics may be inherited with equal frequency by men and women.

Several well-known disorders are transmitted by autosomal recessive genes, including albinism (the absence of pigment in skin, hair, and eyes), cystic fibrosis (an often fatal disorder of the exocrine glands—the secretors of sweat, mucus, saliva, and tears—affecting one of every 1000–3700 white children), and phenylketonuria (one of the so-called inborn errors of metabolism, which can result in permanent mental retardation if not treated soon after birth by manipulation of the infant's diet). Since recessive genes may be carried along for generations without causing the recessive characteristic to appear, they may seem to come 'out of the blue' when at last they manifest themselves in a homozygotic individual. In some recessive diseases, such as phenylketonuria, it is possible by means of medical tests to detect the presence of the damaging recessive gene in an unaffected carrier. This is possible because some recessive genes cause measurable effect on the biochemistry of the body even though they do not cause any outwardly observable trait. Tests for carrier detection, when they are available, greatly simplify the task of estimating risks of recurrence of a recessive illness for purposes of genetic counselling.

Sex-linked dominant

Sex-linkage almost always refers to hereditary transmission by genes located on the X chromosome. The Y chromosome seems to carry very few genes except those that specifically function in the production of male traits during the early development of the male foetus. The only other known trait which has been clearly shown to be Y-linked is the growth of hair around the rims of the ears!

Sex-linked (X-linked) dominant inheritance is quite rare, although it has been suggested, among other theories, as one possible mode of transmission for mood disorders. Since a male transmits one Y chromosome to his son and one X chromosome to his daughter, a male with an X-linked dominant trait cannot transmit the trait to his son, but all of his daughters will be affected. The X-linked dominant gene, when transmitted by the mother, behaves exactly like an autosomal dominant. A mother who has a dominant gene on one of her two X chromosomes will be able to transmit it to both sons and daughters; half of her sons and half of her daughters will inherit the gene, and all who receive it will inherit the dominant trait. Since father-to-son transmission of an X-linked dominant gene is impossible, X-linked dominant characteristics are inherited by twice as many females as males.

Sex-linked recessiveness

Sex-linked recessive traits follow a distinctive pattern of transmission that makes them appear to skip generations before they recur. The classic example of a sex-linked recessive condition is haemophilia A, commonly called the 'bleeding disease' because its victims lack a certain clotting factor in their blood, causing them to bleed excessively either spontaneously or from cuts and bruises. Haemophilia has acquired celebrity because it occurred throughout the royal families of Europe, after it was introduced to them by the descendants of Queen Victoria, who appears to have been a carrier for the disease.

Haemophilia and other sex-linked recessive diseases appear to skip generations because a female inheriting the recessive gene from either her father or mother usually possesses a dominant allele on her other X chromosome to overrule the recessive gene's effects. Thus, she becomes an invisible carrier of the gene, and her own health is not impaired. When she bears children, she is likely to pass the damaging gene to half of her sons and half of her daughters. The daughters who receive it will in all likelihood be unaffected carriers like the mother. But any son who inherits the damaging gene from her will inherit the disease, since his Y chromosome does not contain a locus for that gene, and therefore the action of the recessive gene will not be suppressed. A father who possesses the X-linked recessive characteristic will obviously not be able to pass it on to his son, since his son will receive a Y rather than a X chromosome from him. But all of his daughters will be unaffected carriers of the gene, unless by some remote chance the daughter also receives the recessive X-linked gene from her mother, in which case she would be homozygotic for the recessive gene and would become affected.

Non-Mendelian patterns of heredity

Mendelian models of inheritance describe the most simple and straightforward transmission of genetic traits. The classic models are built on the assumption that genes and chromosomes always follow their own rules and that nothing, including the activity of other genes, the activity of the body's many systems and organs, and influence from the environment, places a roadblock between the action of the gene and its eventual expression as an observable trait. These assumptions are met in the inheritance of many human traits— normal as well as abnormal. But in many other cases the assumptions are violated. The result is a pattern of inheritance that cannot be accounted for by any neat Mendelian scheme.

Genes, as we have already mentioned, break their own rules when they undergo mutation, though mutations are rare. Chromosomes may exchange segments by crossing over during meiosis, causing genes normally inherited together to go their separate hereditary ways. Cross-over is a harmless accident in the life of the chromosome, but other chromosomal aberrations can be severely damaging. Sometimes a developing egg cell accidentally receives two of chromosome number 21; if that egg is fertilized, it will receive another chromosome 21 from the sperm cell. Thus, the zygote and all the cells descending from it have chromosome 21 in triplicate rather than the normal homologous pair. The unfortunate result is Down's syndrome in the child (Down's syndrome used to be called 'mongolism' because the facial features of the typically affected child were erroneously identified with facial character- istics of the Mongolian race). Studies show that late maternal age at conception (after 35) and exposure to excessive radiation (for example, from too many X-rays of the abdomen) significantly increase the risk of such chromosomal accidents. This type of Down's syndrome (trisomy 21), by far the most common, is not hereditary. A rarer, hereditary type of Down's syndrome can be caused by *translocation*, when a piece of one of the chromosomes in the group numbered 13–15 has a piece of chromosome 21–22. Other chromosome translocations also occur, causing the most severe damage when one of the detached pieces with its load of genes is lost within the cell.

Some gene mutations and chromosomal accidents are so damaging that the foetus in which they occur cannot survive until birth. Nature may cover up such an error by causing a spontaneous miscarriage. But because of miscarriages, some serious genetic defects may go unnoticed, and the expected Mendelian ratios of other classically inherited traits may be distorted.

Genetic accidents are not the only causes of non-Mendelian ratios in human beings. Some genes normally exhibit what is called *intermediate inheritance*. This refers to single dominant gene trans- mission where, contrary to the classic model for autosomal dominant inheritance, all homozygotes but only a smaller percentage of heter- ozygotes manifest the dominant trait. Other genes are called *sex- limited* because they behave as dominant genes in one sex but as recessive in the other. So-called 'male-pattern baldness', which appears almost exclusively in men, is an example.

The body has a multitude of biochemical pathways, forming a network of interdependent processes, some of them keyed to the age the individual or his stage of growth, some of them keyed to the individual's diet and other environmental influences, and some keyed to the primary actions of genes. The classic ratios of hereditary traits can be disrupted if at some point in this infinitely complex network a biochemical pathway should become blocked. This phenomenon is called *incomplete penetrance*. One gene, for example, may act like a dominant gene by overruling another's effects, even though the two are not alleles of the same gene and do not appear at the same locus on homologous chromosomes. Another gene, dominant or recessive, may act unimpeded at the primary level, but if the body fails to produce a necessary enzyme, if one of its organs malfunctions, or if infectious disease somehow alters the body's metabolic pathways, the gene's secondary or tertiary effects might be obstructed, with the result that no trait is finally manifested. Other genes show *age-dependent penetrance*, not seeming to become activated until many years after birth. The autosomal dominant gene which causes Huntington's disease, for example, typically does not cause the disease to appear until the victim is between 30 and 55 years old.

Among mental disorders, the patterns of genetic transmission are often obscured by *genetic heterogeneity*. This term refers to the fact that, since the primary causes of almost all mental disorders are unknown, it may easily happen that two or more genetically different disorders having similar symptoms will be placed together under the same diagnostic label. The result, like when your television set receives two or more channels simultaneously, is an unclear picture.

Finally, the same gene may affect two different individuals to different degrees. This is called *variable expressivity*. In one member of a family, a gene may cause a severe disorder, while another member of the family who has received the same gene will develop only a mild condition bordering on normality. Again, one member of a family may develop one subtype of a disorder while another member develops a different but related subtype. The various possible forms of expression of a gene must be taken into account whenever the pattern of transmission of a genetic trait is assessed in a family.

Multiple-gene inheritance

The major cause for non-Mendelian patterns of transmission is *multifactorial inheritance*, which may be considered another major model of transmission alongside classic single-gene heredity. In multifactorial inheritance, the trait is produced by the concurrent actions of many minor genes, triggered or further modified by environmental influences.

When many genes are essential contributors to one trait, the mode of transmission is called *polygenic*, and the individual genes are then called *polygenes*. None of the polygenes by itself would be sufficient to produce the trait in question, but acting in concert they give the affected individual a genetic predisposition which, when activated by the necessary conditions in the environment, manifests as an inherited trait.

Many mental disorders appear to be transmitted polygenically. This type of inheritance results in generally lower overall risks than in classic single-gene heredity. The risk of a polygenic mental disorder to a relative of an affected person rises as the closeness of the relation, the severity of the illness, or the number of other affected relatives increases.

Polygenic inheritance is discussed in greater detail in Chapter 4 under the mode of transmission of schizophrenia.

Methods of genetic research in psychiatry

Genetic researchers studying hereditary components in mental illnesses use mainly three types of investigation. In increasing order of precision they are the *family study*, the *twin study*, and the *adoption study*.

Family studies.

The family study is a basic tool for gathering information from the genetically close relatives of persons who have been diagnosed as having a given mental disorder. Researchers start by identifying a large group of patients all with the same diagnosis, as well as a large group of persons without psychiatric symptoms for the sake of comparison. The relatives of those subjects are identified from medical reports or other registers. Investigators then gather the information they need by searching through case records and, preferably, by personally interviewing the family members. This is a relatively easy type of investigation, which tells the investigators what kinds of mental disorders are found in various classes of relatives, and how the relatives of psychiatric patients compare with members of the general population.

A properly conducted family study can show whether a mental disorder tends to run in families and whether any other mental illnesses or other disorders also occur with increased frequency in the relatives. When the genetic mode of transmission of a given mental disorder is unknown, exact risk estimates are impossible to calculate for family members, but with information gathered in family studies, researchers can draw up *empirical risk estimates*, i.e. practical estimates based on observations of the frequency of given disorders among various classes of relatives.

It is important to bear in mind the limitations of family studies as tools for genetic research. They may show that an illness tends to run in families, and a familial tendency is a strong hint of underlying genetic influence. But 'familial' is not synonymous with 'genetic'. Hereditary and environmental influences may both contribute to the familial tendency. The family study is not designed to prove whether or not a genetic influence is present.

Twin studies

Twin studies are specifically designed to distinguish the influence of genetic factors in given traits. The idea behind the twin study is that two people with identical genes—if such people could be found—should be expected to have identical traits, if those traits are purely genetic in origin. Of course, pairs of people with identical genes can be found: they are called *identical twins*. Such twins, also called *monozygotic*, develop simultaneously with their partners from a single fertilized egg (or 'zygote'); thus, monozygotic twin partners have completely identical sets of genes. If one twin possesses a genetic trait, the identical partner should possess it too—similarity between partners for a given trait is called *intrapair concordance*. The intrapair concordance among monozygotic twins for strictly genetic traits is expected to be 100 per cent.

Beside monozygotic twins, however, there is a second type called *fraternal* or *dizygotic* twins. Dizygotic twins develop simultaneously in the womb from two separate zygotes; thus, dizygotic partners can be expected to have only half of their genes in common, just like ordinary brothers and sisters.

The twin study capitalizes on the innate difference between monozygotic and dizygotic twin types by investigating whether for a given trait the monozygotic concordance rate is significantly higher than the dizygotic concordance rate. A higher monozygotic rate is taken as strong evidence that the disorder in question has a genetic basis.

Methods of genetic research in psychiatry

It cannot be assumed, however, that all concordance between twin partners is caused by underlying genetic factors; intrapair concordance might be encouraged by the sharing of an environment as well as by genetic similarity. Therefore, the twin study is not entirely able to distinguish environmental from genetic influences in given traits. Some researchers have attempted to overcome this limitation by studying monozygotic twins who have been separated from each other at birth and raised in different homes. This design cancels out the possible contributions of shared home environment to intrapair concordance rates. Unfortunately, however, monozygotic twin pairs raised apart are so rare that it is practically impossible to gather a large sample for study.

Adoption studies

Adoption studies are some of the best available methods for teasing apart the tangled threads of environmental and hereditary influences in mental disorders. The placement of infants in adoptive homes shortly after their birth defines three groups suitable for study: the adopted children themselves when they are grown, the biological parents, and the adoptive parents. This three-way comparison is very instructive, since the biological parents are related to the adopted child genetically without sharing the same environment, while the adoptive parents share the same home environment with the adopted child without being genetically related. Significant similarity between the adopted child and the biological parents indicates hereditary influence, while similarity between the adopted child and the adoptive parents must be due either to chance or to the influences of a shared environment—it cannot be due to heredity.

Researchers sometimes begin by choosing a sample of adopted children who have become mentally ill and then investigate whether an increased rate of that illness can be found in either the biological or the adoptive parents. Such a study may be expanded to include all the close relatives of the adopted child, both biological and adoptive. At other times researchers begin by choosing a sample of mentally ill parents who gave their children up for adoption at birth; the researchers then investigate whether the adopted-away children develop the same illness as their parents at any higher rate than adopted children whose biological parents were normal. When the adopted children develop the same disorder as their biological parents at a significantly increased rate despite having no physical contact or communication with them, the evidence strongly indicates that a genetic component is present in the disorder. Another design, called a *cross-fostering* study, is used to determine whether rearing by a mentally disordered parent in itself, without the existence of a genetic relationship between parent and child, is correlated with a higher risk of the parent's disorder in the child. For this purpose, researchers select persons who were born to normal parents but were adopted away at birth into the home of a parent who later became mentally ill. They follow up the adopted children into adulthood; if a significantly increased rate of disorder is found among the cross-fostered subjects, the increase has to be attributed to environmental influences. An absence of the disorder among the cross-fostered subjects does not prove that the disorder is genetic, but it does indicate that rearing by a disordered parent is not sufficient by itself to cause the disorder in a child.

As later chapters in this book report, family, twin, and adoption studies have established the existence of genetic factors in many of the major psychiatric illnesses. Psychiatrists still do not know the precise causes of mental illnesses or the precise modes of their inheritance, but the evidence of genetic factors at work in these disorders has opened up challenging new avenues of genetic, biochemical, and psychosocial research. Such research holds out hope that sometime in the not too distant future the root causes of mental illnesses will be discovered and that from such discoveries more successful treatments and possibly cures will be developed.

3
Psychiatric genetic counselling

Psychiatric genetic research meets everyday life in the confidential interaction between an expert who can communicate the available facts about the inheritance of mental disorders and a client who needs to know them. This interaction is called psychiatric genetic counselling.

The purpose of such counselling, broadly defined, is to help alleviate any human problem related to the occurrence or possibility of recurrence of a genetically influenced mental disorder within a family.

In practical terms, this may mean that a psychiatric genetic counsellor will help a young married couple in their struggle to decide whether they should have children of their own when a serious mental illness appears to run in one of their families. Or as often happens, the counsellor may relieve parents' anxiety about the future health of children already born, simply by providing reliable information and helping them to understand it. Or the counsellor may clarify the diagnosis of a mental disorder by organizing and interpreting information about the family's psychiatric history.

Is genetic counselling in psychiatry legitimate?

Many people, including some professionals in the mental health sciences and services, have resisted the concept of genetic counselling for psychiatric disorders. One reason, perhaps, is that after a scientifically premature swell of enthusiasm for genetic theories of mental disorder early in this century, the world witnessed the abhorrent spectacle of pseudogenetic theories being used as tools of repression and genocide before and during the Second World War. Even within respectable medical establishments during the first half of this century, the enthusiasm for 'cleaning up the gene pool' through eugenics programs sometimes carried overtones of class and racial prejudice. Today, some people continue to regard genetic counselling with suspicion, as if it still campaigned under the banner of eugenics.

Furthermore, many psychiatrists, doctors, psychologists, social workers, counsellors, and educated laypeople have formed their opinions about the origins of mental illness during the years when psychiatry was heavily under the sway of psychological theories of behaviour. Psychological theories dominated research, clinical practice, and education in psychiatry for the middle years of this century, fostering attitudes of either scepticism or indifference toward genetic theories among workers in the medical and mental health professions. These attitudes still linger in some professional circles.

Finally, some people have objected to genetic counselling for psychiatric disorders on the grounds that no one yet knows the true causes of mental illnesses. How, they challenge, can psychiatric genetic counselling do any good—or more strongly, how can it keep from doing irreparable harm—so long as psychiatry remains in the dark about whether, how, and which genes are at work in mental illnesses? If researchers are not sure what gene defects are involved in psychiatric disorders, how can counsellors estimate familial risks with any accuracy?

Despite these objections, there are many reasons why psychiatric genetic counselling today is not only a legitimate enterprise but also an important and increasingly necessary one.

The chief reason is that adoption studies, reinforced by the results of twin and family studies, have convincingly shown that there is a strong genetic component in the more heavily studied psychiatric syndromes such as schizophrenia, manic depression, presenile dementias, and alcoholism. Counsellors now have a dependable body of evidence to pass on to psychiatric patients and their worried relatives. Of course, such knowlege is no adequate substitute for a clear understanding of how a disease is caused, what precise genetic pattern is present, and how the damaging genes do their work. But we have only to look at a familiar medical disease to find a useful parallel. One expert on medical and psychiatric genetics, Dr James Neel, has called diabetes 'a geneticist's nightmare' because, just as in schizophrenia, the underlying causes of the disease are unknown and the disease probably includes several subtypes within it with differing degrees of genetic influence. Yet patients with diabetes and their relatives routinely ask for and successfully receive genetic counselling

Is genetic counselling in psychiatry legitimate?

If counsellors knew the precise causes of mental disorders, their job would be considerably easier. But genetic counselling is possible even without such information. Instead of using exact risk estimates, based on knowledge of the precise mechanism of inheritance, the counsellor may use figures called empirical risk estimates: percentages drawn from the pooled results of large family studies which show the rate at which a given mental disorder occurs in each of the different classes of relatives of an affected person. For most psychiatric disorders, these are the best risk estimates available, and though they must be handled with appropriate caution, they serve an important function in genetic counselling for psychiatric disorders.

Another justification for counselling is the simple fact that mentally ill patients and their relatives often worry, about themselves, about each other, and about the children they may bring into the world. Such worry is secondary to the kind of suffering directly caused by mental illness, but through genetic counselling some of it may be relieved.

Who should counsel?

Psychiatric genetic counselling must begin with an accurate diagnosis of the disorder in question. Because it is so important to pinpoint the diagnosis, the ideal counsellor is probably a psychiatrist who is thoroughly familiar with the pitfalls of psychiatric diagnosis and is in close contact with other experts—neurologists, paediatricians, geneticists, and many kinds of medical technicians—who have the special skills sometimes needed to establish a correct diagnosis. The specialist in behavioural genetics with training in psychiatry may also be able to provide dependable psychiatric genetic counselling.

The *International directory of genetics services* (5th edition, 1977) lists 564 centres in the world (292 in the USA) where genetic counselling is offered. Only 78 of these (34 USA) are specifically listed as being able to provide psychiatric genetic counselling. These are for the most part large medical centres and university teaching hospitals where many of the necessary services are within easy reach.

Undoubtedly, there are many psychiatrists who are also equipped for counselling in centres not listed in the *International directory*. No one has a precise count of these or a list of their locations.

Clearly, however, there are not enough psychiatrists in the world at present to meet the full need for genetic counselling services. True, many people have no idea that genetic counselling is available for psychiatric disorders, so that the demand is not overwhelming. But if every family who needed such counselling would seek it out, the psychiatric profession would not be able to endure the load.

For better or worse, then, much of the psychiatric genetic counselling that is done will be done informally by family doctors or other doctors and counsellors without special training in either genetics or psychiatric diagnosis. Family physicians may be suited for counselling in some ways, since they may have had more intimate contact with family members, perhaps over several generations, than may be possible for a psychiatrist. The danger, obviously, is that much counselling will be done by doctors and counsellors who are not yet adequately informed about the latest developments in psychiatric genetics, or even the classic ones. The shortage of properly trained psychiatric counsellors therefore gives urgency to the education of health care professionals in this area.

Who needs psychiatric genetic counselling?
Anyone who has reason to worry about the recurrence of mental disorders in their family may benefit from genetic counselling. However, different members of a family will approach genetic counselling from different standpoints. The typical concerns of a psychiatric patient are not always the same as those of the patient's relatives or of prospective spouses.

Whether or not psychiatric patients become involved in genetic counselling often depends on the nature of the illness. In some disorders such as presenile dementias, the patient's memory and judgement are often damaged by the disease so that he or she cannot process the information a genetic counsellor has to provide responsibly. Victims of Huntington's disease have been known to lack any sense of the hereditary danger to their offspring, even after genetic counselling in which they learned about the 50 per cent risk of transmitting the disease to their children. Chronic schizophrenics often lack insight into their own illness and fail to comprehend or appreciate the questions at stake in genetic counselling. In addition, their symptoms may include thought and communication disorders which block meaningful exchange with the counsellor.

Where psychiatric patients keep their awarenesss and general

40

level of intellectual ability, as in many cases of manic depression and alcoholism, patients are especially likely to need and want gentic counselling.

Family members come for counselling when they see either a few especially severe mental disorders or many less severe ones in their family. Unaffected family members want to know if they will receive the disorder because of their relation to the patient. Relatives just beginning to build their own families want to know whether they can safely have children of their own. If they have already had children, they may want to know if the children will become affected in the future. They may need advice on how and when to inform the children of their at-risk status. Sometimes relatives are also concerned on behalf of the patient's children who have come into their custody.

In cases of manic depression, it often happens that patients and their whole families come in together for counselling. This arrangement can simplify the drawing up of family psychiatric histories and can assure that everyone in the family shares the same understanding of the nature of the problem and the options for dealing with it.

When families seek counselling as a group, they do not usually come spontaneously at the first sign of a mental illness in one of the family members. On the contrary, before counselling they may have experienced generations of financial, emotional, occupational, legal, and social turmoil because of the illness in their family. These may have left deep scars and caused them to erect emotional barriers as defences against further pain. Family corporate interests may vie with personal interests or individuals in the family. Such problems are often compounded by a collective sense of guilt. All of these factors complicate the counselling process. At the same time, genetic counselling may offer an opportunity for some long-buried anxieties within these families to be aired and overcome.

One of the tragedies about genetic disorders is that precautions are so often considered too late, when a serious disease that might have been avoided occurs and does irreparable harm. When prospective spouses volunteer for counselling, they are in an excellent position to prevent that sort of tragedy by educating themselves properly. A prospective spouse will seek counselling when the partner's family or his/her own contains serious cases of mental illness. The partner related to a mentally ill person may be worried about his own liability, and both partners together may wonder about the prospects for their potential children. Anxiety about such matters sometimes interferes with the partners' relationship, threatening the impending marriage before it begins. Thus, genetic counselling may have a useful dimension as marriage counselling, affecting all future decisions about childbearing and rearing.

It is important to stress that the purpose of counselling is not necessarily to warn the prospective couple about the danger of passing on a disease. Many couples have an exaggerated fear of possible consequences and need to be convinced that their fears are unwarranted.

Seven stages of psychiatric genetic counselling

Since counselling addresses so many different concerns and has to be adapted to people of such widely varying capabilities, there is no way to describe an average consultation. Every patient and family has unique concerns and problems; the counsellor must find the unique set of responses that best takes into account each client's dilemma, temperament, intellect, and emotional states, within the limits of the general goals of psychiatric genetic counselling.

Notwithstanding the uniqueness of each consultation, thorough genetic counselling typically includes some form of each of the following seven stages: (1) diagnosis, (2) taking the family history, (3) establishing the risk of recurrence, (4) evaluating the needs of the client, (5) weighing burden, risk, and benefit, (6) forming a plan of action, and (7) follow-up.

Diagnosis

Getting an accurate diagnosis is the critically important first step in counselling. It is far more dangerous to proceed with genetic counselling when the diagnosis is mistaken than when the hereditary mechanism is unknown.

Accurate diagnosis means one that is based not only on the whole picture of clinical symptoms but also on what is called a 'longitudinal' view of the illness, i.e. the total course and progress of the disease over time, as well as a complete family history.

The standard diagnostic manual used in the United States as of January 1980, is the *Diagnostic and statistical manual*, 3rd edition (DSM-III), published by the American Psychiatric Association. Throughout the world, a broader system of classification called the *International classification of diseases*, 9th revision, *Clinical modifi-*

cation (ICD-9-CM) has been in effect since January 1979. ICD-9-CM is a modification of the original ICD-9 published by the World Health Organization in 1977.

For research purposes, diagnostic lines must be drawn as precisely as possible. Therefore, more strictly defined criteria have been devised for some of the major psychiatric disorders and their subtypes (Feighner *et al.* 1972; Tsuang and Winokur 1974); however, such criteria are not available for all diagnostic categories. The counsellor must use all available resources, including other psychiatrists, special laboratory services, and his or her own clinical experience to make an accurate diagnosis before proceeding with counselling.

Numerous problems complicate the matter of diagnosis in psychiatry. Sets of diagnostic criteria have their own histories, and despite everyone's wish to have one set of criteria agreeable to all, differences of interpretation have arisen. Variations in diagnostic criteria thus may exist from region to region and from clinic to clinic within one region.

Another problem is that special expertise required for making an accurate diagnosis may be out of reach of both the patient and the genetic counsellor.

Problems may arise because of the deceptiveness of psychiatric symptoms. Several disorders having identical symptoms may actually be separate diseases, some inheritable, some only partially inheritable, and others uninheritable. That is to say, a type of mental disorder defined only on the basis of symptoms may be genetically heterogeneous.

These difficulties highlight the importance of seeking out an expert (who in most cases will be a psychiatrist) to obtain an accurate diagnosis. Counselling based on a misdiagnosis is at best a waste of time, but at worst it could be seriously misleading.

Taking the family history
This stage is not exactly a psychiatric version of *Roots*; the object of taking a family psychiatric history is not to dig up records of the most distant relatives but to note the frequency and pattern of mental illnesses in the genetically more important nearer relatives.

To take a family history, the counsellor needs records and reliable personal recollections to establish the number and order of relatives, as well as such information as parental age, ethnic background, possible occurrence of abortions, stillbirths, miscarriages, and deaths, and the ages, sex, and health of living brothers, sisters, and children. The counsellor pays primary attention to disorders that have occurred in the immediate family (that is, among 'first-degree' relatives), but the family history must also record the health of other blood relatives, such as grandparents, uncles, aunts, and cousins.

If the diagnosis is uncertain before the family history is drawn up, the sketch of familial disease patterns may contain information to clear up the uncertainty. In certain larger families, a recognizable Mendelian pattern of inheritance may appear, such as a sex-linked recessive pattern that appears to skip generations, or, in the case of Huntington's disease, an autosomal dominant pattern. Such a pattern would give the counsellor a means of calculating exact risk figures for the patient or the relatives.

Establishing the risk of recurrence
After the family history has been filled in, the counsellor can proceed to estimate, if possible, the likelihood that the patient's illness will reappear in a member of his family. If the counsellor knows the exact genetic mode of inheritance of the disorder, he can calculate an exact risk estimate. For example, if the disorder is transmitted by an autosomal dominant gene, as in Huntington's disease, the exact risk to each child of the affected person is 50 per cent—one chance out of two—modified downward according to the number of years the offspring has passed through the normal period of highest risk without being affected. The method of calculating such figures is described in greater detail in Chapter 6.

If the mode of inheritance is not known, as is the case with most mental disorders, the counsellor can turn to empirical risk estimates, i.e. estimates of likelihood based on studies of the frequency of an illness in various classes of relatives in affected families.

As one might expect, empirical estimates are less accurate than exact risks. But given their limits, empirical estimates can be very informative to clients. In using them, however, both counsellor and client must keep an important qualification in mind: the empirical risk estimate refers to the likelihood that the person in question will be affected *some time during his or her lifetime*. The estimate would best describe the risk to an unborn baby. But most mental disorders

44

occur many years after birth, during a period of years recognized as the period of highest risk. The period differs for each specific mental illness. Schizophrenia, for example, usually appears between late adolescence and approximately age 40. An unmodified empirical estimate will be too high for those who have partially or entirely passed through the standard risk period without being affected.

Drs Edmund Murphy and Gary Chase of the Johns Hopkins University School of Medicine, authors of a major textbook on genetic counselling (1975), have categorized the three basic types of information available to a counsellor for determining an accurate risk estimate.

1. 'Empirical information' consists of estimates based on data collected in controlled research projects. Such studies may provide the counsellor with information about the rate of certain gene mutations in the population, the average age of onset of a disorder, or the estimated risk of recurrence in first-degree relatives (parents, brothers and sisters, or children).

2. 'Modular information' is based on a theoretical understanding of the 'mode' of inheritance of a certain trait or complex of traits.

3. 'Particular information' is all of the data contained in the family history of the person or persons whose genes are in question. Murphy and Chase further divide this category into 'posterior information' (i.e. information about all persons in the family whose genetic makeup could be influenced by the genes of the ill family member—children, grandchildren, etc.) and 'anterior information' (i.e. all information on members of the pedigree who could not possibly have been influenced by the patient's genes, namely parents, uncles and aunts, nephews and nieces, step-children, and so on).

Each of these types of information must be interpreted cautiously so that they do not become confused with one another.

Evaluating the needs of the client
From the client's first contact with the counsellor, a process of mutual evaluation is begun. The client wants to be reassured of the counsellor's intelligence, competence, compassion, confidence, candour, and willingness to explore all dimensions of the problem. The client has a right to expect these things of a counsellor and a right to look elsewhere if the counsellor does not satisfy these

expectations.

The counsellor, in turn, can be expected to evaluate his clients, not in the sense of judging their personal worth but in the sense of judging their ability and readiness to assimilate the type of information which he may be called upon to communicate to them. Before he can go on to help the clients to formulate reasonable attitudes and courses of action, he must determine whether the clients are able to act reasonably at that time. In this process, several questions have to be answered.

What is the relation between the client and the person who has the mental illness? Is the mental health of living children in question? What is their relation to the mentally-ill person? Is future childbearing a concern? These questions are necessary to pinpoint the individual whose genetic endowment is in question.

Other questions concern the ability of the client to participate in meaningful discussion and rational decision-making. What is the client's own subjective estimate of the risk to offspring? Does the client fully understand the meaning of the diagnosis? Has the client responded emotionally to it so as to block out other information? Do tensions within the family prevent the client from thinking clearly and autonomously? What is the client's general level of intelligence? What is his or her grasp of genetic principles? What is the client's attitude toward having children?

Ignoring these questions, or underestimating their importance, could lead to a breakdown of communication at some point in the counselling process.

Burden, risk, and benefit in the balance

Some clients need counselling to help them decide whether or not they should have children. For them the next stage of counselling will be critically important, for they are called upon, with the counsellor's help, to weigh the multitude of variables affecting them, with an eye toward coming to a decision. Much depends on the client's ability to remain objective in spite of popular stigmas, previous misconceptions, and the attraction of psychological escape mechanisms.

The complex conceptual processes involved at this stage can be simplified if one views them as essentially a balancing act involving three elements, which we shall call *burden*, *risk*, and *benefit*. Let us picture *burden* and *benefit* as boxes at opposite ends of a seesaw,

Seven stages of psychiatric genetic counselling

able to change in size and weight according to the client's perception of their importance or seriousness. The seesaw will pivot upon a fulcrum labelled *risk*, which can be moved to the left or right, with rightward movement representing *increase* of risk. Notice that when the risk is increased, the leverage (or relative weight) of *burden* increases.

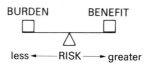

Before altering the values of the variables of the illustration, we must define the three terms.

Burden is a word used by genetic counsellors to refer to the total amount of strain and expense caused by an illness over the course of the patient's lifetime. The greatest cost, of course, is to the affected individual, but relatives may consciously bear a heavier burden from the illness than the affected individual himself.

It is impossible to measure the precise burden of a mental disorder. All burdens are ultimately carried in the mind, and as yet we have no way of measuring precisely what the mind knows and feels. The question of burden finally becomes entangled with unknowable factors of willpower, hope, self-confidence, honesty, and many other intangibles.

Yet we can improve our appreciation of the burden of mental illnesses on human lives by singling out some of the dimensions of life they affect. Their most obvious and easily measurable impact is financial. Mental illnesses mean visits to clinics, doctors' bills, payments for treatments and medicines, loss of wages, fees for special nursing or community facilities, public tax money consumed, and much more.

But the financial cost is only the visible tip of the iceberg. How can one calculate the weight of lost human potential or human resources consumed in minimizing loss? How can one measure physical and emotional pain? The cost of a suicide? What is the lifelong impact of childlessness? How much does embarrassment weigh, or grief over lost opportunities? Who can reckon the damage of emotional distress or guilt? And if all of these costs are extended over time, how great does the burden grow?

When an illness passes quickly and does little to disrupt normal patterns of life, the burden is, of course, very low. Again, a mild disorder, even though it may last a lifetime, will not impose an insufferable burden on a family. But neither will a severe disorder that quickly abates or is so severe that it soon causes the death of the affected person. But when illness destroys a life slowly and consumes human and financial resources, say, over 30 or 40 years, the burden is very high.

Many of the mental disorders discussed in this book can be considered high burden disorders. Chronic schizophrenia, severe alcoholism, presenile dementias—all are relatively long-term disorders calling for somewhat elaborate professional intervention. With the arrival of effective drug treatments, manic depression is changing for some patients from a high burden illness to intermediate or even low burden. Many of the personality disorders and neuroses about which psychiatry has very little genetic information are of the 'long-term nonsevere' variety.

If burden was hard to define, the term *benefit* is even more elusive. *Benefit* refers to the satisfaction, advantage, or fulfilment parents expect to receive from bringing a child into the world. Certainly this is something which parents themselves are undecided about; they may not understand the depth of their feelings about procreation until the crisis which precipitates psychiatric genetic counselling forces them to examine the question. The relative weight of *benefit* in our equation can be influenced by many variables: religious convictions, political ideology, number of children already born in the family, family traditions, social status, the state of the national economy, the symbolic value attached to childbearing, and many other factors.

Most of the literature on genetic counselling is silent on the concept of *benefit*. Yet in recent years an increasing number of studies have appeared which claim that a surprising number of people who receive genetic counselling choose to have more children even when they have been warned of ominously high recurrence risks. These results might not be so surprising if we had a better understanding of the many factors impelling parents towards bearing children.

The concept of *risk* has already been discussed. The most important concept here is the distinction between exact risks and empirical risks. Risk rates are usually communicated to clients in the form of percentages. A 75 per cent risk means that a person has three chances out of four of becoming ill at some time during the span of a life—not gambling odds! An 8 per cent risk would mean that the individual in question has eight chances out of 100. This estimate may seem low in comparison to the previous one, and yet for a given disease it may represent a tenfold increase over the risk for that disease in the general population.

The model (Figure 6) may be set in operation, but with this reservation: it is deliberately made very simple and strictly logical—people, on the other hand, are neither.

The seesaw in the horizontal position shown in the first diagram represents a situation where the parents have reached stalemate.

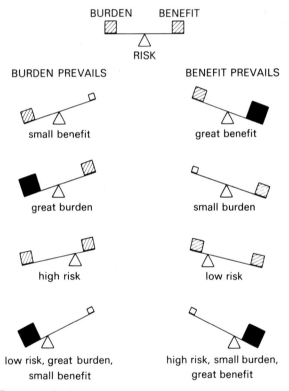

Fig. 6. The seesaw in the horizontal position represents parents whose fear of recurrence of the disease exactly balances their desire for a child and the risk is neither high or low enough to tip their decision in either direction. In each figure, the values are changed, illustrating disorders of greater and lesser burden, raised or lowered estimates of risk, and greater and lesser desires for bearing children.

Their fear of a potential recurrence of the disease exactly balances their desire for a child, and the element of risk is neither high nor low enough to tip their decision in either direction. In each figure, the values are changed, illustrating disorders of greater and lesser burden, raised or lowered estimates of risk, and greater and lesser desire for bearing children.

The figures in the left-hand column here represent situations in which the fear of burden would prevail in the decision-making process. The right-hand column represents situations in which the anticipated benefit of childbearing would prevail. Infinitely many variations might occur in real-life situations. Obviously, a model like this will not make a decision for anyone. But it may be useful to some clients as a kind of mirror, helping them to analyse their own feelings.

Forming a plan of action

After a psychiatric genetic counsellor has helped his clients to understand for themselves the relative values of burden, risk, and benefit in their own situation, he must help them decide upon an informed course of action that is properly their own. It is not enough for the counsellor to churn out facts like an adding machine and leave all the wrestling with indeterminates and anxiety to the clients. On the other hand, if the counsellor becomes too directive, he may encroach on the human prerogatives and legal rights of his clients. Based on his clinical and counselling experience, he may have a clearer understanding of the consequences of certain courses of action, and the clients may value his recommendations or cautions. In the end, however, the counsellor must commit all responsibility for the decision to the parents, without pressuring or rushing them.

There are many obstacles to surmount on the path towards responsible action. First of all, though one may speak of burden, risk, and benefit as hard quantities, the fact remains that each is at best a guess. Perceived burden, perceived risk, and perceived benefit are the true variables in the scales of decision-making. And it must be conceded that the counsellor's own appraisals of burden, risk, and benefit, informed though they may be, are nevertheless personal perceptions, too.

Because the outcome of psychiatric genetic counselling will always be determined more by perceptions of facts than by facts in themselves, we ought to note some of the ways in which distorted perceptions are able to skew the decision-making process.

Genetic counsellors often report that their clients claim to be interested in gauging only the burden of an illness while ignoring risk estimates. This is an understandable reaction in light of the sometimes complicated arithmetic and unfamiliar genetic theories which may scramble the picture for those people (the majority) who are not specially educated in these areas. The danger of such an approach, however, is twofold. First, the client who considers only burden, because he ignores at least half of the pertinent information available to him, evades reality. Secondly, behind this apparently one-sided process of thought, the process of decision-making inevitably goes on unawares. Instead of using a reliable measure of risk, however, the client will fall back on private institutions likely to be founded on prejudices and misinformation. Instead of considering openly the benefits of childbearing as he or she perceives them, the client may be guided by emotional reflexes or old wives' tales. The decision-making process can be better controlled and altered through education if the elements in it are exposed to the light.

In genetic counselling clients are subject to many kinds of distortion in their perceptions of risk. Powerful fears may incline a person to overestimate the risk to himself or his offspring. Other clients falsely simplify the whole concept of risk. One researcher has reported a patient who used this pretty piece of logic: 'I'll either get the disease or I won't. So my chance is 50–50, or one out of two.' As it turned out, he was far above the mark.

Other clients cut short the whole process by imagining that the disease is bound to recur, then considering whether the disease would be bearable or not. Such 'worst case scenarios' may be valuable if they help a person achieve a measure of readiness, but they ignore relevant probabilities in favour of general possibilities.

As mentioned elsewhere in this book, risk figures themselves may contain hidden pitfalls. Some are calculated from studies using narrow diagnostic criteria for selecting a sample. The results of these studies strictly apply one to those who have the same narrowly defined illness. Again, lifetime risk figures must be modified as the unaffected person passes safely through the standard period of risk. The actual risk may also rise or fall in accordance with the number and nearness of affected relatives and with the severity of the familial disorder. Finally, empirical risk figures apply generally to broad classes of relatives, but the figures may not be accurate in large families where a specific pattern of inheritance is detectable.

Follow-up

The last stage of psychiatric genetic counselling is the follow-up, which may be done in person, by mail, or by whatever method is most effective. The counsellor wants to be assured whether the most crucial information has been retained. Follow-up gives the clients a chance to ask new questions that have arisen in the interim since the counselling session and to discuss continuing problems not solved in the original exchange. Follow-up may also uncover new data that might be useful for revising a risk estimate or determining the exact mode of transmission of a disorder.

A professional response to need

Psychiatric genetic counselling must not be thought of as the dispensing of information for the curious. It is a professional response to the needs of people who live with daily suffering brought about by the hereditary nature of mental illnesses. These people live in a culture that tries to ignore mental illness, stigmatizes it unfairly, and permits misinformation to abound. Often patients themselves and their relatives perpetuate the sense of stigma.

This kind of suffering is assuaged in genetic counselling when the counsellor, in an emotionally supportive atmosphere, provides enlightening information based on genetic, biological, and epidemiological research in psychiatry and helps his clients to deal with it. Such research can already answer many pressing questions such as whether mental illnesses are hereditary, what distinct varieties of mental illness exist, and whether mental illnesses have a biological basis.

The answers emerging from research point to a much closer resemblance between medical diseases and mental illnesses than either the public or the medical professions have previously realized. The research that will unlock the deepest mysteries of mental illness still lies ahead. But as new discoveries are made, society also needs improved education, both within the health care professions and in the community, to erase the stigma on mental illnesses. Psychiatric genetic counselling is in the front line of that effort.

PLATE 1: Schizophrenic withdrawal and isolation
(courtesy of Sid Bernstein, Rockland Research Centre)

PLATE 2: A case of Huntington's chorea showing
marionette-like movements and facial grimaces

4

Schizophrenia

This chapter and the following three take the reader on a guided tour through genetic studies of schizophrenia, manic depression, presenile dementias, and alcoholism. According to the most recently published statistics from the US Department of Health (1977), these illnesses, with the exception of presenile dementia, account for by far the greatest number of annual admissions to mental hospitals in the United States. All four are seriously debilitating disorders which 'run in families'. As evidence mounts showing that each of these disorders has a strong biological component which can be genetically transmitted from parents to their children, victims of these disorders and their relatives have an increasing need for genetic counselling.

Schizophrenia ranks first in this discussion for several reasons. Of all the mental illnesses responsible for suffering in society, schizophrenia probably causes more lengthy hospitalizations, more chaos in family life, more exorbitant costs to individuals and governments, and more fear than any other. Because it is such an outstanding threat to life and happiness and because its causes are a centuries-old puzzle, it has been studied more than any other mental disease.

What is schizophrenia?

Contrary to what many people think, schizophrenia is not the same illness as split or multiple personality. Cases of individuals having up to thirteen distinct personalities make for splashy headlines and gripping TV serials, but such disorders are exceedingly rare. Schizophrenia, on the other hand, is relatively common, with a risk of affecting more than one out of every 150 people in the general population. Rather than being a splitting of a mind into many separate personalities, it is the splitting apart of basic mental functions (speaking, moving, feeling) within a single personality. We will illustrate the disorder with a case history of a patient who possesses many of the most serious features of schizophrenia.

Case history

Twenty-six-year-old Janet Douglas (fictitious name) was brought to a mental health centre by her father. Upon admission she told the examining doctor she did not think she needed help. After further questions, she made the odd claim, with a straight face, that all the people in her city could hear what she was thinking. It all started five years ago, she said, when the President of the United States ordered the FBI to plant 'truth serum' in her drinking water. Suddenly, she broke into an uncontrolled giggle, wrinkling up her face, rolling her head, and saying, almost incomprehensibly: 'But I fool them . . . the way they are. My eyes can speak of the beauty. I say love-words and pattern words I've found out until everybody quits the way I make them. . . .' She giggled wildly into her cupped hand. Shortly afterwards, she told the doctor that many times when she 'starts to be perfect' she hears the voices of neighbours in the air. They talk about her 'sins' and usually punish her by taking thoughts out of her mind, leaving her powerless to think. These voices frighten her so much when they come that they deprive her of sleep and meals for as many as four days at a time. While she described her torments, however, her giggling faded and she seemed to become inappropriately calm, attaching no emotional depth to her own words, as if she were reciting a grocery list. Probed by the doctor for details about the strange voices, she fell silent again, a crooked grin spreading across her face with no apparent pleasure behind it, and she could only respond in single words to simple questions like 'What day is it today?'

By her father's account, she has not always been like this. She had anyone's definition of a normal childhood, although she had few very close friends, and went on to college, majoring in chemistry before dropping out. She worked at many jobs but in the middle of her college years began to lose her ability to concentrate. The quality of her work dropped off over a year's time. She took so many unexcused absences that she was fired three times before she finally stopped working altogether. At the same time, she lost interest in sewing, her favourite hobby, and ignored her former friends. She became irritable with members of her family, offering implausible and fragmentary reports of her activities and seeming to attach an incoherent private symbolism to the words and gestures of other people. The giggling, strange delusions, odd mannerisms, and broken manner of speaking had not begun to occur until about a half year before her current admission to the hospital.

The diagnostic workup included a complete physical examination and laboratory report, which discovered no physical illness or injury, no organic process at work in her brain, and no history of drug abuse which could satisfactorily explain her symptoms or the earlier course of her illness.

Based on all of his observations, the examining doctor reached a diagnosis of 'Schizophrenia, Hebephrenic Subtype'.

What is schizophrenia?

Will Janet Douglas and other patients with symptoms like hers ever be well again? Unfortunately, the presence of severe psychotic symptoms like Janet's (hearing voices, paranoid delusions, disjointed thoughts, fragmented speech, diminution of the emotional life, and so on) typically declares a schizophrenic illness that will leave the victim chronically impaired, despite efforts at treatment or rehabilitation. However, a certain percentage of those with schizophrenic symptoms tend to recover from their illness over time.

Studies of schizophrenic patients who recover have found that several features are consistently associated with a good prognosis. These include the following: a family history unmarked by other cases of schizophrenia or marked by manic depression in relatives, a quick onset of schizophrenic symptoms (over less than six months) without a prior process of deterioration, or the presence of a stressful experience that clearly seemed to have precipitated the illness.

The signs usually associated with a poor prognosis are onset of illness early in life (i.e. in middle or late adolescence), a long pattern of social maladjustment or eccentric behaviour before the actual flareup of schizophrenic symptoms, and a general pattern of steady deterioration in mental functions after that. Schizophrenics who fit this description are likely to have more cases of schizophrenia among their relatives than 'good prognosis' schizophrenics.

Is schizophrenia inheritable?

Evidence from genetic research

Ever since the turn of the century, when schizophrenia was first acknowledged as a distinct illness, doctors have observed that new cases tend to occur in families where other cases of schizophrenia are already present. The tendency of this disease to run in families was a hint that it might have an underlying genetic cause.

The familial pattern alone, however, proved nothing. Many diseases appear in successive generations within a family without being genetically transmitted. Consider a family of coal miners in which grandfather, father, and son all have developed black lung disease by early middle age. Their continual inhalation of coal dust in the mines sufficiently explains their disease; there is no need to raise a hypothesis of genetic transmission. Other diseases can mimic

55

a genetic mode of transmission when, for instance, parents pass on harmful dietary habits, viruses, or other infectious agents to their children. Schizophrenia itself could conceivably be transmitted by a certain kind of virus. None has ever been discovered, but as yet psychiatrists do not have enough information to rule out the possibility.

Thus, even though many family studies have confirmed the tendency of schizophrenia to run in families, thereby pointing strongly toward a genetic mode of inheritance, they by themselves do not prove that genes are active causes of the disease.

The case for a strong genetic component in schizophrenia is built primarily on the results of two other kinds of studies: investigations of schizophrenia in twins, and adoption studies, which investigate patterns of illness in adopted persons, their biological relatives, and their adoptive families.

Twin studies of schizophrenia

A twin study, as explained in Chapter 2, compares the concordance rate of identical (monozygotic) twin pairs against the concordance rate of fraternal (dizygotic) twin pairs for a specific disease. Since identical-twin partners have identical sets of genes, their concordance rate for a genetic disease should be higher than the rate for fraternal twins (whose genes are only half-shared). When the increase of concordance rate in identical-twin pairs over fraternal pairs is too great to be explained by chance alone, the disorder is assumed to have an underlying genetic component. One hundred per cent concordance among identical-twin pairs would indicate that the disease is entirely transmitted by genes.

Psychiatric geneticists have carried out twin studies of schizophrenia all over the world. When the results are placed in one large data pool, they yield representative concordance rates for schizophrenia in identical and fraternal twins. If schizophrenia is genetically transmitted, a statistically significant increase should appear in the concordance for schizophrenia among identical-twin pairs. And that, indeed, is what the figures show. A pooled concordance rate for schizophrenia in identical-twin pairs is 45.6 per cent, compared to 13.7 per cent in the fraternal pairs. (Reminder: these figures do not mean that 45.6 per cent of all identical twins have schizophrenia but that in 45.6 per cent of the identical-twin pairs where one partner had schizophrenia, the other partner had it too.)

Is schizophrenia inheritable?

Researchers have refined the twin study method by distinguishing between same-sex and opposite-sex fraternal pairs, then comparing these groups separately against identical-twin pairs. The rationale for this methodological twist is that since same-sex fraternal-twin partners have similar sex chromosomes, they should be genetically more similar than opposite-sex fraternal partners, while still not genetically as similar within pairs as identical-twin partners (who are always of the same sex). Thus, if schizophrenia is genetically transmitted, the concordance rates for same-sex fraternal-twin pairs ought to fall somewhere between the corresponding rates for opposite-sex fraternal-twin pairs and identical-twin pairs.

The expected pattern can be seen in the pooled data summarized by Shields and Slater (1967), where the concordance rate for schizophrenia rises from 5.6 per cent to 12 per cent to 57.5 per cent for opposite-sex fraternal twins, same-sex fraternal twins, and identical-twin pairs, respectively.

Some researchers have objected to twin studies of schizophrenia on the grounds that being raised as a twin may create a unique combination of stresses, leading to confusion of identity in the partners and inadequate formation of their egos, conditions which could lead to schizophrenia, according to some theories. Identical twins could be assumed to suffer more of these stresses than fraternal twins. Such a hypothesis would help to explain the higher concordance rates for schizophrenia among identical-twin pairs.

However, if being reared as an identical twin significantly enhances the likelihood of one's becoming schizophrenic, we would expect to find disproportionately many twins in random samplings of schizophrenics. That has never been proved to be the case.

Twin studies indicated in a wholesale manner that genetic factors are at work in schizophrenia. But where does the action of genes end and the influence of environmental conditions begin? Twin studies are not designed to pick apart the tangled strands of environmental and genetic influences in a disorder because they study persons who have been raised in the same family milieu with their twin partners. And yet the figures on identical-twin concordance for schizophrenia leave plenty of room to infer the influence of nongenetic factors. Less than half of the identical-twin pairs in the pooled studies were concordant for schizophrenia, despite the fact that, as a rule, the partners in each pair had identical sets of genes.

Schizophrenia

Obviously, genes are not the only factors influencing schizophrenia. And yet, elevated rates of identical-twin concordance for schizophrenia cannot be satisfactorily explained by environmental factors either. A few rare studies have investigated concordance for schizophrenia in identical-twin pairs separated from each other at birth; in such cases, any potentially harmful effect of being reared in the environment of an identical-twin partner is cancelled out. Eight such investigations in seven countries were summarized by Gottesman and Shields (1972). Out of 17 identical-twin pairs reared apart, 11 (64 per cent) were concordant for schizophrenia. The high rate of concordance for schizophrenia in genetically identical siblings, even when they had no exposure to one another, has to be explained, in part, as the result of genetic factors.

Adoption studies of schizophrenia

In the early 1960s, several teams of investigators in the United States and Europe undertook adoption studies of schizophrenia for the first time. The adoption study in any of its various designs nullifies the possiblity of postnatal environmental interaction between the adopted child and his or her biological relatives. This control mechanism makes the adoption study one of the best available means for separating the relative contributions of environment and heredity in schizophrenia.

The earliest adoption study of schizophrenia was carried out in Oregon by Dr Leonard Heston (1966), who investigated 47 adopted children whose natural mothers were schizophrenic. They were matched for sex, type of foster placement, and length of time spent in childcare institutions with a control group of 50 adopted children whose biological mothers were not schizophrenic. The researcher followed up on the adopted children to a mean age of 36. Although the sample was small, a clear pattern emerged: the children of schizophrenia mothers had a significantly higher rate of mental illness than the controls, not only of schizophrenia (11 per cent against 0.0 per cent) but also of neurotic personality (28 per cent against 14 per cent), antisocial personality (19 per cent against 4 per cent), and mental deficiency (9 per cent against 0.0 per cent).

These results suggested to the researcher that an array of psychiatric disorders besides typical schizophrenia might have been transmitted via the schizophrenic mothers. If the definition of the inherited trait could be broadened to include the whole array, the researcher proposed, the rate of transmission would approach 50 per cent, a figure suggestively in line with a theory of single-gene dominant transmission. The fly in the ointment, however, was that the investigation had not taken into account the fathers of the adopted children. If the schizophrenic mothers had tended to mate with men who had inheritable psychiatric disorders of their own, some of the extra disorders in the adopted children could have been accounted for through them.

The Oregon study of adopted children showed that the children of a schizophrenic parent develop a higher than average rate of schizophrenia even when removed from their parents at or soon after birth. Since adopted children with normal biological parents had a much lower rate of psychiatric illness, the increase in the children of schizophrenic mothers could not be blamed on some damaging effect of living with adoptive parents.

More definitive evidence of a genetic component in schizophrenia came from a series of adoption studies in Copenhagen, conducted by American doctors Seymour Kety, David Rosenthal, and Paul Wender, with their Danish colleagues Drs Fini Shulsinger, Joseph Welner, Bjorn Jacobisen, and Lise Ostergaard. Denmark was chosen as a site for the study because the Danish adoption and psychiatric records, to which the psychiatrists had access, are some of the best in the world. These studies have become landmarks in the study of genetic factors in schizophrenia.

The major study in the series (Kety *et al.* 1975) was an investigation of the families of 33 adopted persons who had become schizophrenic. Its purpose was to determine on the basis of personal interviews whether persons biologically related to a person with schizophrenia would be distinguished from adoptive relatives by a higher rate of schizophrenia. 'Schizophrenia' in this study was defined as a disorder falling within the spectrum of schizophrenic disorders (typical schizophrenia, borderline schizophrenia, reactive schizophrenia, or 'doubtful' schizophrenia).

The schizophrenic adopted people were carefully matched with 33 normal adopted people. In all, then, four groups of relatives were able to be designated for the study: biological and adoptive relatives of both schizophrenic and normal adopted people. One of these four groups (the biological relatives of schizophrenic adopted people) stood apart from the other three in being genetically related to schizophrenics who were removed from them at birth. This group alone was found to have a markedly higher rate of mental illness. Overall, 21 per cent of the biological relatives of the adopted schizophrenics fell within the schizophrenia spectrum, compared with 11 per cent of the biological relatives of controls. The difference is statistically very significant.

This study included an especially interesting subgroup, the paternal half-brothers and half-sisters of the adopted people. Normal siblings have half of their genes in common, while half-siblings, who have only one parent in common, share 25 per cent of their genes, still a significant genetic relationship. But in a study of paternal half-siblings of adopted people, the influence of environmental factors is even more effectively controlled than in an ordinary adoption study. Not only do the adopted children and their paternal half-siblings grow up in different homes, but, since they are related only through a father, no adopted child spends even the first nine months of life in the same womb as his or her paternal half-siblings. Thus the possibility of disease being transmitted somehow through factors in the prenatal or neonatal environment can be controlled. Any significant increase of mental illness in the paternal half-siblings must be due to genetic factors transmitted from the fathers (unless the fathers tend to choose mates with transmissable psychiatric illnesses, a hypothesis that needs to be tested).

The researchers investigated paternal half-siblings separately and found that of those related to schizophrenics, 8 (13 per cent) were either chronic or borderline schizophrenics, and 6 (10 per cent) were diagnosed as having doubtful schizophrenia. Each diagnosis was given to only 1 (1.6 per cent) of the paternal half-siblings of the control group.

Two other studies in the Danish series further certified the presence of genetic influences in schizophrenia. One by Rosenthal *et al.* (1968) found that children of schizophrenic natural parents adopted into the homes of non-schizophrenic foster parents at an average age of six months, still developed schizophrenia (i.e. schizophrenia spectrum disorders) at a much higher rate than adopted-away children of normal parents. One-third of the adopted-away offspring of schizophrenic parents became schizophrenic, compared to only about one-fifth of the control group. The former group included three cases of chronic schizophrenia while the latter group had none.

Can the type of rearing provided by adopting parents cause

schizophrenia in their adopted children? A further study shed some light on this question by using a 'cross-fostering' technique. Wender *et al.* (1974) controlled the genetic factor in their study by choosing two groups of subjects born to normal parents; one group had been adopted into the homes of parents who later became schizophrenic. The other group had been adopted into normal homes. For comparison, a third group was added, which consisted of people born to schizophrenic parents but adopted soon after birth into the homes of normal parents.

If rearing by schizophrenic parents had been a significant cause of schizophrenia in the children and if gene action had not been an important influence, one would have expected the cross-fostered group to have the highest rate of schizophrenia of all three groups in the study. But, in fact, the results ran exactly counter to this hypothesis. The cross-fostered group—those born to normal parents but raised by genetically unrelated schizophrenic parents—had the lowest rate of schizophrenia at 4.8 per cent (the investigators attributed the slightly higher rate in the control group of problems with drawing a random sample). The highest rating, 19.7 per cent, went to the group genetically related to schizophrenics but raised by normal adoptive parents.

The investigators concluded that rearing by a schizophrenic parent is not a significant cause of schizophrenia in the child if the child is not already genetically predisposed to the disease. Of course, the study did not completely rule out certain so-called 'schizophrenogenic' patterns of rearing by non-schizophrenic parents or other environmental influences as potential contributing factors in some other forms of schizophrenia, but it strengthens the case that genes are of major importance in the transmission of typical schizophrenia.

Adoption studies have their limitations, as the Danish investigators have repeatedly pointed out. They do not explain the biochemical actions of genes underlying schizophrenia. They do not tell us whether one gene or many are at work. Finally, they do not completely separate genetic from environmental variables, as Dr Kety warned in a keynote address at the First International Symposium of Immunological Components in Schizophrenia:

> Even an adopted child has spent 9 months in the uterus of his biologic mother and has received a certain amount of mothering at her hands. It is not impossible that during that time the mother could have transmitted to

the fetus some nongenetic biologic or psychosocial factor that might bring about schizophrenia 15 years later.

What sort of factor might be transmitted which could cause such delayed effects? One candidate might be a virus behaving somewhat like the herpes type, which could be transmitted through the male genitourinary tract at conception and lie dormant in the affected individual for 15 years or more before being triggered by a combination of biological and psychosocial conditions. No such virus has yet been discovered, but none of the data emerging from family studies, twin studies, or adoption studies completely overrules the possibility that one might exist.

It might be recalled, however, that the Danish study of paternal half-siblings of schizophrenic adopted people controlled the maternal influences on the prenatal environment of those who later developed schizophrenia. For that reason, Dr Kety said, he believes that the results of the half-sibling study are the most compelling evidence we have for the operation of genetic factors in schizophrenia.

Cumulatively, the results of family, twin, and adoption studies of schizophrenia yield strong evidence that genes play a major causative role in many cases of schizophrenia. The twin studies also point up the important role of non-hereditary factors by finding less than 50 per cent concordance for schizophrenia among identical twins. Adoption studies cast no doubt on the importance of those hypothetical environmental factors; as yet, however, no specific conditions of rearing, diet, infection, or other psychosocial stress have been pinpointed as necessary or sufficient causes for schizophrenia. The cross-fostering study by Wender *et al.* does not discount all supposed environmental contributions to schizophrenia; it merely cancels out one of them, namely, rearing by schizophrenic parents, as a possible cause of schizophrenia in persons who have not been endowed with genes for the disease.

How is schizophrenia inherited?

Let us assume (reasonably, in view of the research just discussed) that schizophrenia is genetically inherited. What is the mode of its transmission? By 'mode of transmission' we mean the functional arrangement of the genes in transmitting the disorder, whether single or multiple locus, dominant or recessive, autosomal or sex-linked, and so on.

How is schizophrenia inherited?

If schizophrenia fitted a classic (Mendelian) model of inheritance, as described in Chapter 2, it could be expected to follow certain predictable patterns within families. For example, if it was a dominant trait with 100 per cent penetrance, it would appear in about half the children of a schizophrenic parent; only afflicted parents would have afflicted children (since none would be unaffected carriers); and the disease would never appear to skip a generation. But no such clear-cut pattern is seen in schizophrenia, nor is any other simple Mendelian pattern. How, then, is schizophrenia inherited?

The models of inheritance proposed for schizophrenia fall into three general classes: (1) single-gene (monogenic), (2) multiple gene (polygenic), and (3) multiple mode (genetic heterogeneity). We cannot cover all the ramifications of these theories here, since they call for more sophistication in genetics than this book presumes of the reader. Furthermore, the various theories are still contending with each other to such an extent that they have little bearing yet on the practical problems of genetic counselling. What follows is necessarily a brief summary of the main theories proposed.

Monogenic theories

A monogenic theory of schizophrenia, as the name implies, proposes that the disease is produced by a single gene which causes schizophrenia when it is expressed. One version of the monogenic theory holds that the gene will manifest itself only in homozygotes, i.e. those who have the damaging allele at the same locus on both members of a chromosome pair; another version posits a gene that will be expressed in either homozygotes or heterozygotes, i.e. when the allele is present on at least one chromosome in a pair. These two theories correspond to Mendelian autosomal recessive and dominant transmission, respectively. The problem facing monogenic theories is to account for the failure of the gene to affect individuals at the frequency expected under a Mendelian model of inheritance.

Very few researchers support a recessive monogenic model. For a recessive gene to be expressed, both parents have to contribute the recessive gene to their child. That child will then be homozygotic for that gene. If both parents are unaffected carriers, one out of four of their children should be homozygotic, affected, and one out of two will be unaffected carriers. Since the damaging gene is quite uncommon in the population, the chance that any of these children will

63

meet and marry another unaffected carrier as their parents had done is very much less frequent. Therefore, the children of a person affected by a recessive trait are much less likely to inherit the trait than that person's brothers and sisters. But this expectation is not met in schizophrenia; the children of schizophrenic parents are just as frequently affected as the parents' siblings.

Most monogenic models propose that a dominant gene is at work but that its effect is somehow reduced. One version of the theory accounts for the reduction of effect as being the result of other genes at other loci on the chromosome or at loci on other chromosomes. Such genes might chemically blockade the schizophrenic gene when present in the right strengths or combinations. Another dominant gene theory (Slater 1958) suggests that the dominant gene is transmitted faithfully according to the Mendelian pattern but that in a certain percentage of cases its effects are altered, obscured, or blocked among the countless chemical interchanges that take place in the human developmental network. You may recall that dominant gene theory of Heston and his colleagues was of this nature. Many genes for other human traits have 'reduced penetrance' of this sort. Various researchers have estimated the rate of penetrance at anywhere from 6 per cent to 13 per cent. A third dominant gene theory posits 'intermediate effects', where schizophrenia is always manifested in homozygotes but only in about one out of four heterozygotes.

Polygenic theories

Polygenic or multifactorial theories argue that schizophrenia is caused by a pile-up of the effects of many genes, none strong enough by itself to cause schizophrenia, but sufficiently strong to cause the disorder when acting in concert with each other and with influences from the environment. Polygenic theories are relatively new on the psychiatric scene and are still being statistically developed and tested.

One polygenic theory is called the 'continuous phenotypic variation theory', a mouthful of words meaning that many different genes, each making a small and approximately equal contribution to the development of schizophrenia, are spread around in the population, forming a normal distribution of continuously varied traits ('phenotypes') among affected individuals. Figure 7 shows what a graph of such a distribution might look like. At one extreme of the distribution (on the left of the diagram) a minority of affected individuals develop

How is schizophrenia inherited?

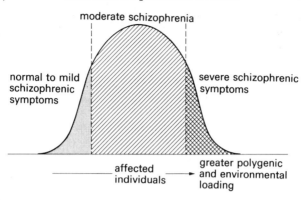

Fig. 7.

schizophrenic traits of a mild sort, sometimes indistinguishable from normal behaviour. The majority of affected individuals cluster around an average, having a moderate dose of polygenes and show noticeable schizophrenic symptoms, some of them bordering on serious, long-term disorders. Those with the greatest load of polygenes, coupled with strong pressures from the environment, fall at the extreme right end of the continuum. According to this model, typical schizophrenia does not differ in kind but only in degree from borderline schizophrenia, schizoid personality, latent schizophrenia, or even from what is called normality.

A second polygenic theory called the 'quasi-continuous variation theory' proposes that schizophrenia is caused by many genes, some with a powerful effect and some with a weak effect, none of which is sufficient to produce schizophrenia by itself but which may take effect when certain combinations are present in an individual, subject to environmental influences. This model assumes that a certain trait which cannot be observed, called a liability (or vulnerability) to schizophrenia, is continuously distributed throughout the population in a humping curve that resembles the distribution of ordinary traits like height or IQ (see Figure 8). According to this theory, some in the general population would have a liability surpassing a certain threshold, and they would manifest the disease. Others who have sub-threshold liability would remain unaffected. If we were to investigate the relatives of those who surpass the threshold, we should find that they, as a group, have a higher liability than members of the general population. This is reflected in the lower diagram by the rightward displacement of

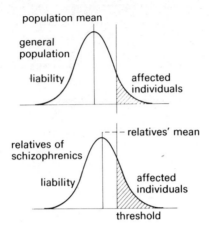

Fig. 8.

the liability curve; notice that the shaded area under the curve and to the right of the threshold becomes larger when the mean liability of the relatives is increased.

This model is very general, since it has only one threshold. It assumes that the underlying cause of schizophrenia is multifactorial and that the disease is an all-or-nothing state: one either has it or does not. That might be an adequate general description of simple traits like cleft palate, but in our previous discussions we have already seen that schizophrenia is probably not so simple as all that. There may be mild forms distinct from severe forms, early onset forms distinct from late onset forms, subtypes characterized by one set of symtoms distinct from subtypes characterized by a different set of symptoms, and many other distinctions as well within the syndrome generally called schizophrenia.

The quasi-continuous variation model can be made more flexible to reflect these distinctions by having two or more thresholds added. For example, let us assume that schizophrenia has two distinct thresholds. One is T_1, which distinguishes normal health from a mild form of schizophrenia and another is T_2 which distinguishes mild schizophrenia from a severe form. Figure 9 illustrates how the proportion of relatives affected with a mild or severe schizophrenia might change as their mean liability changes, according to whether they are relatives of severe or mild schizophrenics or

66

How is schizophrenia inherited?

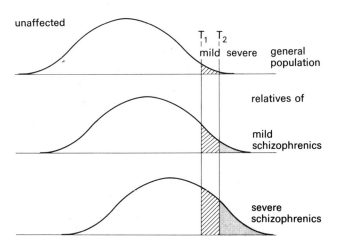

FIG. 9.

members of the general population. Other thresholds might be added by breaking up schizophrenia into categories based on sex, type of clinical symptoms, course of illness, and so on. To test whether the model actually fits the patterns of schizophrenia observed in relatives, biomathematicians can calculate values for a series of 'parameters' (features such as the prevalence of mild or severe forms in the general population, the frequency with which a certain allele occurs in the general population, the frequency with which a certain genotype manifests a certain phenotype, etc.). From these values they can estimate how frequently the disease should occur among the relatives of an affected person if the assumptions of the model are true. Then they can compare these estimates against the actual data collected in large family studies, using standard statistical tests to tell them whether the model is a 'good fit' or not. In this way, some models can be rejected and others allowed. Theoretically, sophisticated versions of the multi-factorial threshold models should be able to distinguish between single major locus and polygenic modes of inheritance. But that hope is still in psychiatry's future. The models work best when applied to pure subgroups of schizophrenia using large volumes of data collected from family studies. So far, however, the ideal statistical requirements have never been met; no conclusions have been reached concerning the validity of the quasi-continuous variation model of inheritance in schizophrenia.

Often both a dominant gene model and a polygenic model will fit the same family data well enough so that neither can be ruled out. In such cases, undoubtedly, both the psychiatric genetic counsellor and the counsulting family would appreciate a clearer understanding of how schizophrenia is inherited, and yet either model may provide useful risk estimates.

Genetic heterogeneity

Monogenic and polygenic theories assume that schizophrenia is a single disease, while using different devices to explain why this hypothetically unitary disease is not inherited along classic hereditary lines. Genetic heterogeneity theories, on the other hand, suggest that what is usually lumped under one name—schizophrenia—is actually an Irish stew of separate diseases, each type caused by its own specific gene or combination thereof and having its own distinct mode of transmission, but all disguised under one broth of clinical symptoms. Most researchers in psychiatric genetics now agree that schizophrenia is not a unitary disorder, although they by no means agree how it should be properly subdivided.

Figure 10 shows how the class of diseases called schizophrenia may be broken down into subcategories, based on current knowledge about age and sex distribution, clinical symptoms, presence of triggering life-experiences, patterns of transmission within families, existence of genetic factors, and many other kinds of information. In the diagram the solid connecting lines signify divisions which have been established past reasonable doubt by clinical, family, and genetic studies. The broken lines indicate subdivisions which may be correct on the basis of current evidence but which need to be verified by further research.

The first major division is between 'True' and 'Organic' forms of schizophrenia. Organic forms are those which can be clearly traced to physical causes such as brain damage, drug abuse, or infectious diseases. In such cases, the symptoms of schizophrenia are secondary to the underlying physical problems. True schizophrenias, by contrast, are those which have no discoverable physical causes—although as our knowledge of the body's chemistry expands, more and more 'true' schizophrenias may be found to have underlying physical causes.

How is schizophrenia inherited?

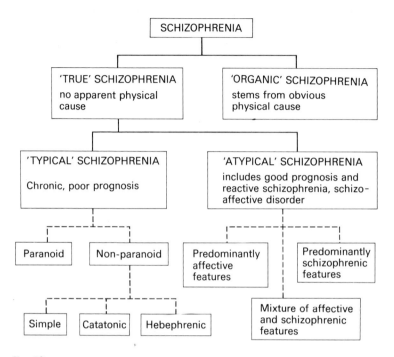

Fig. 10.

Within the family of true schizophrenias we can distinguish between typical and atypical forms. Atypical schizophrenias are those which deviate significantly from the chronic, deteriorating course or severe clinical symptoms which characterize typical schizophrenias. The atypical category includes schizophrenic disorders which occur as reactions to major environmental or psychosocial stresses, those whose features promise eventual recovery, those which lack some of the primary symptoms of typical schizophrenia, and those which are mixtures of schizophrenic and manic-depressive symptoms.

These divisions are widely accepted in psychiatry. Some controversy attends the subdivisions of typical schizophrenia at present. It has been suggested on the basis of clinical and genetic studies that paranoid and non-paranoid forms of schizophrenia should be treated as distinct subtypes. The non-paranoid group in turn can possibly be broken up into three subtypes called simple, catatonic, and hebephrenic. The evidence for these subdivisions is their tendency to breed true within families and to differ from one another in the frequency of schizophrenia among relatives. However, the available evidence allows much room for debate about this system of classification.

Atypical schizophrenia may also be subclassifiable, as follow-up and family studies of atypical schizophrenics have shown. Some atypical schizophrenics have a disorder which may resemble manic depression more than schizophrenia; other forms of atypical schizophrenia may belong with schizophrenia; and a third subtype, belonging neither with schizophrenia nor with mood disorders, but incorporating symptoms of both, may be a distinct disease. These subdivisions also need to be tested by further research.

What are the risks to relatives of schizophrenics?

Exact risk estimates will not be available in psychiatric genetic counselling so long as the mode of transmission of schizophrenia remains undiscovered. But a genetic counsellor can bridge the gap in medical knowledge by using empirical estimates taken from pools of family studies which determine the frequency of schizophrenia among various classes of relatives of schizophrenic parents.

Table 1 shows the lifetime risks of schizophrenia to various relatives of someone with the classic form of the disease. Since these percentages are simple observations of frequency, they leave unresolved all questions about the relative contributions of environment and heredity. But since environmental contributions to schizophrenia are still as little understood as genetic ones, the possibility of bias is a moot point.

At the bottom of the table you see that the risk for members of the general population is 0.86 per cent, according to a pool of 19 studies in 6 countries. This figure, translated, means that at some time in their lives, 86 out of 10 000 people randomly chosen from the general population will become schizophrenic. The risks for relatives of schizophrenics should be compared against this basic rate.

Note that the risks for brothers and sisters are about ten times as high as those for the general population. The risks for children of schizophrenics, at 12.3 per cent, are nearly 15 times the population rate.

The risks to second-degree relatives—uncles and aunts, nephews and nieces, grandchildren, and half-siblings—are roughly three

What are the risks to relatives of schizophrenics?

times the population rate, though they are considerably lower than
the risks to relatives in the immediate family circle.

TABLE 1. *Risks to relatives of schizophrenics**

Relation	Risk (%)
First-degree relatives	
Parents	4.4
Brothers and sisters	8.5
neither parent schizophrenic	8.2
one parent schizophrenic	13.8
Fraternal twin, opposite sex**	5.6
same sex**	12.0
Identical twin**	57.7
Children	12.3
both parents schizophrenic	36.6
Second-degree relatives	
Uncles and aunts	2.0
Nephews and nieces	2.2
Grandchildren	2.8
Half-brothers/sisters	3.2
First cousins (third-degree relatives)	2.9
General population	0.86

*Unless otherwise noted, figures are based on Slater and Cowie (1971); data
mainly derived from pooled data of Zerbin-Rudin (1967), with only cases of definite
schizophrenia counted.
**Pooled data based on Shields and Slater (1967).

If schizophrenia is a genetically transmitted disease, we would
expect to see three patterns in the family risk figures. First, we
would expect to see higher rates of schizophrenia among relatives
than among the general population. This is obviously seen. Secondly,
we would expect higher risks among near relatives than among
distant ones. This pattern, too, appears in the table. In fact, the
highest figure listed here is 57.7 per cent, the risk to an identical
twin, who would have a set of genes exactly similar to his schizophrenic twin partner's. Thirdly, we would expect the risks to rise as
the number of near relatives with schizophrenia increases. Again,
matching the prediction, the table shows that the risk to brothers
and sisters rises from 8.2 per cent to 13.8 per cent when one of the

parents is also schizophrenic; even more striking is the leap in risk to nearly 40 per cent for children of two schizophrenic parents.

These patterns do not prove a genetic hypothesis, but they are fully consistent with it.

Putting risk figures in perspective

A list of percentages may be misleading, and potentially harmful, unless interpreted correctly. Counsellors will help their clients by maintaining a proper amount of both respect and scepticism toward such tables. The numbers are not chiselled in stone. Relatives of schizophrenic patients need to read what is written between the lines. In many cases, clients are influenced not so much by the risk figures themselves but by a subjective sense of their value, which may be distorted by fear of personally becoming ill or by inaccurate weighing of the burden of schizophrenia. To the hypercautious, a risk of 5 per cent may spell certain disaster if it is ten times the normal rate. To the cavalier, a 40 per cent risk may seem like fair odds.

Counsellor and client together must decide whether risk figures like those cited in this chapter are relevant to the specific family situation and whether they should be adjusted upward or downward to reflect particular patterns of illness among the relatives.

The figures in the table are based mostly on European studies which have typically used stricter definitions of schizophrenia for sample selection than American studies. Study samples are limited to those who suffered serious, non-remitting disorders. However, many cases coming to the attention of a genetic counsellor will be milder disorders falling within the schizophrenia spectrum. Unfortunately, no good risk figures are yet available for milder schizophrenic disorders, though in all likelihood the rates of schizophrenia in relatives of individuals with such disorders are lower than those listed in the table. On the other hand, the rate of mild schizophrenia and other psychiatric disorders among the relatives of typical schizophrenics may be higher than the rate of typical schizophrenia. A few studies have shown that for relatives of some patients with quick onset forms of schizophrenia, the risk of manic depression may be greater than the risk of schizophrenia, even though the patient's major symptoms are schizophrenic.

Some of the estimates given in the table may have to be lowered for any of the following reasons. If the patient has an organic type of schizophrenia, the risks to his or her relatives are probably not much higher than those for the general population, unless there are other cases of true schizophrenia in the family. Some relatives with high genetic liability to schizophrenia may be partly or completely through the years of highest risk at the time of counselling without having developed the disorder. Their risk is lessened accordingly, though not entirely eliminated.

Researchers have also found that the children of paranoid schizophrenics have only about half the risk of becoming psychotic as the children of non-paranoid schizophrenics. This may reflect the fact that paranoid schizophrenia characteristically occurs later in life than other subtypes, and researchers have noticed a correlation between late onset of schizophrenia and lower risks of schizophrenia among the relatives.

It is also a favourable sign if the family contains very few other affected persons beside the patient or only very mild cases.

The risk will increase according to the number of relatives with definite schizophrenia and the nearness of their relation to the person in question. It also increases if the patient has one of the earliest onset varieties of schizophrenia such as hebephrenia or catatonia. The genetic counsellor should take note of atypical schizophrenic disorders within the family, even though their genetic relationship to typical schizophrenia is not very clearly understood. The presence of many of them in one relation may increase the risk of schizophrenia in the others.

Schizophrenia, especially in its typical forms, is a burdensome disease, destroying minds in the prime of life and consuming many of the precious resources of families and societies. Because it is inheritable, it creates sobering responsibilities for would-be parents whose children are at risk. Unfortunately, little can be done at present to lessen the risks for those most vulnerable. Twin studies permit the conclusion that rearing and environment have a large share in causing schizophrenia but do not single out any specific environmental condition as a culprit. Wender's cross-fostering study demonstrates that rearing by a schizophrenic parent is not sufficient by itself to cause schizophrenia in a genetically normal child. On the other hand, we know that adopting a child away from his schizophrenic parents into the home of normal parents does not get rid of the inherited risk. Hence, in our present imperfect state of knowledge we cannot reduce the risk of schizophrenia for a child of a schizophrenic parent by altering the environment. Nor can we prophesy who is going to develop schizophrenia later in life on the basis of some foolproof behavioural or chemical marker. So we cannot lessen the risk by administering preventive medical treatments. Schizophrenia can at present be treated only after it has declared itself. Then, in most cases, it is not possible to eliminate the burden of the disease through a total cure; rather, the disorder must be controlled by means of antipsychotic drugs—phenothiazines, thioxanthenes, butyrophenones—which have been discovered since the introduction of chlorpromazine in 1952. These drugs reduce the burden of illness by controlling the flagrant symptoms of schizophrenia so that the patient is more responsive to therapy and supportive arrangements in the home or institutional environment.

Summary

Convincing evidence that genetic factors are present in schizophrenia has come from twin and adoption studies around the world. Identical-twin pairs, even when raised separately, show significantly higher concordance for schizophrenia than fraternal twins, thus indicating genetic influence. However, the fact that even identical pairs are concordant less than half the time implies that environmental factors also contribute strongly to the disease. Adoption studies in the United States and Denmark provide the clearest evidence of genetic influence since by design they separate genetic from environmental influences. These studies demonstrated that: (1) the adopted-way children of schizophrenic parents develop a higher rate of schizophrenia than adopted children whose natural parents are normal; (2) biological relatives of schizophrenic adopted children, including paternal half-siblings, show significantly more schizophrenia than biological relatives of normal adopted children; and (3) children born to normal parents but raised by schizophrenic adoptive parents did not show a significantly elevated rate of schizophrenia.

The mode of transmission in schizophrenia is not yet known, although theories of autosomal dominance with reduced penetrance (about 6–13 per cent) and polygenic theories can be supported by available data from family studies. It appears likely that schizophrenia includes several genetically distinct subtypes within it, though no agreement has been reached on how precisely the subtypes should be delineated.

Summary

The lifetime risk of schizophrenia for members of the general population is 0.86 per cent, but for brothers and sisters of typical schizophrenics, this rate becomes ten times higher; children of typical schizophrenics are about fifteen times more likely to become schizophrenic than members of the general population. These figures can be adjusted upwards or downwards to reflect critical factors such as the patient's age at onset of illness, the number of relatives affected, the severity and type of familial cases, and the closeness of the affected relatives to the person(s) for whom psychiatric genetic counselling is sought.

5

Mood disorders:
mania and depression

The purpose of this chapter is to draw together current knowledge about genetic components in mood disorders (commonly called manic depression). The victim of one of these mental disorders typically becomes divorced from reality because of incapacitating mood swings toward extreme euphoria, deep depression, or both in alternation. The chapter concludes with a discussion on the heightened risk of affective disorder among close relatives of affected patients and other questions that arise in genetic counselling.

The studies summarized here help to demystify the illness. They show that mood disorders are quite common in the population, that they are genetically inheritable, and that the symptoms can be controlled in many cases with medications which alter the body's chemistry, though in ways not yet fully understood.

What are mania and depression?

The two words 'manic depression' refer to the two main manifestations of mood disorders at opposite ends of the emotional continuum. Periods of extremely elevated mood are called mania, while periods of extreme unhappiness, despair, or delusions of guilt are called depression.

The words 'mania', 'maniac', 'depression', and the like are often dropped loosely in ordinary conversation, but in the language of psychiatry they have a more precise meaning. They properly refer to mental states, in which the mental disturbance is severe enough to cause gross mental and behavioural abnormalities resulting in serious social maladjustments in the life of the affected individual.

Mania

People who have never experienced or witnessed manic behaviour sometimes find it hard to imagine that anyone might be 'too happy'. But mania is not to be confused with the normal experience of

76

elation such as that which accompanies falling in love, the return of spring, the birth of a baby, etc.

The following case history illustrates the abnormal and damaging effects of a manic psychosis.

Case history

A 34-year-old self-employed man, referred to hereafter as Dave, began to experience a feeling of euphoria, as his mother recounted, for no apparent reason. Family members and friends commented on his extraordinary cheerfulness for several days not suspecting something might be wrong. But over the next few days he became increasingly elated to the point of giddiness.

Anticipating a tax refund from the United States government, he began to brag that he would soon inherit a huge fortune through a private deal with Uncle Sam. He overdrew his bank account on a spending spree for items like a new refrigerator, a motorboat, formal clothes for which he had no use, garden furniture, and other expensive items. He made the rounds of parents and friends trying to borrow large amounts of cash, breezily promising each lender that when his tax refund arrived he would promptly repay them a hundred times over.

His talk during this period became animated to the point of incomprehensibility. Jokes, exclamations, witty asides, and attempts at impersonation piled out on top of each other as if he were trying to out-talk himself or to catch up with his racing mind. Any new stimulus set him off on a tangent. But when others tried to interrupt or calm him down, he would suddenly answer with insults, scolding, and cursing.

At night, after entire evenings spent dashing around in a car and surprising friends with unannounced visits, he would return home, not to sleep but to scribble late into the night in a journal of his poetry and personal thoughts. He claimed, with no basis in fact, that paperback book and movie rights had been bought from him for a screenplay he intended to write. After two weeks of feverish activity, grandiose posturing, sleeplessness, and missed meals, Dave had lost more than ten pounds and was complaining of constant dryness of the mouth. But his expenditure of energy had not tapered off.

Dave was arrested for reckless driving two and a half weeks after the onset of his symptoms. He was caught weaving dangerously on a freeway, his car door wide open and his left leg jutting out perilously close to the roadway, as if he thought he could stop the vehicle by dragging his foot. The arrest resulted in Dave's admission to a mental health centre, where he was given a diagnosis of manic psychosis and started on a programme of lithium therapy, after admission examinations showed no physical disorder or history of drug abuse which could have accounted for his symptoms.

Dave experienced a psychotic form of mania, as distinguished from less severe conditions such as hypomania, characterized by manic symptoms too mild to be considered psychotic. The person affected with hypomania may be overactive, talkative, and abnormally elated but, in spite of all, fairly well adjusted, free of delusions, and able to perform daily routines without serious interruption.

Depression: case history

Depressive symptoms have an opposite appearance from manic elation and grandiose delusions. In the case of a 39-year-old woman we will call Mrs Johnson, mother of three school-aged children, the typical symptoms of a depressive psychosis are clearly evident:

Mrs Johnson had three main interests before the onset of her symptoms: bringing up her children, exercising at a local health spa, and reading historical novels. Without any notable preceding events in her life, she began to behave as if each of these activities was becoming too strenuous to carry on. Soon she could not bring herself to open a book, dropped her membership in the health club, and, though she spent all of her time at home, abandoned most of her duties as a homemaker, complaining of fatigue and feeling overwhelmed with life. Despite her husband's attempts to encourage her to take up her former hobbies, she began to spend whole mornings in bed. When she rose, she would pace agitatedly, often weeping, clutching her face and wringing her hands, claiming that she felt useless and unloved. She was aware that her behaviour was making her children and husband suffer, but this awareness simply compounded her feelings of despondency and guilt. She seized on the idea that everyone she knew would be happier if she would kill herself. After about three weeks of decline, she had almost stopped speaking to people, had become listless, and showed the effects of sleepless nights and poor eating in a thin and haggard appearance. When she did speak, her comments centred on her feelings of being the most worthless creature who had ever lived.

At the end of the third week of symptoms, she agreed with her husband that she should be admitted to the hospital. There she was admitted with a diagnosis of major depressive disorder. She showed modest improvement after about two months, following treatment with electroconvulsive therapy (ECT) coupled with antidepressant medication.

Studies show that both manic and depressive patients have an abnormally high risk of death by unnatural causes. Mrs Johnson dwelled on the idea of suicide but never attempted it. Tragically, many other depressed patients make the attempt and succeed. Manics, on the other hand, are more prone to accidental deaths resulting from their own reckless behaviour and poor judgment.

What are mania and depression?

The unipolar–bipolar distinction

Mania and depression have been regarded as different manifestations of one illness for about 75 years, ever since their symptoms, typical course, and familial characteristics were first clearly described and distinguished from what we now call schizophrenia by the eminent German psychiatrist, Emil Kraepelin.

But mood disorders, cannot be split evenly into pure manic disorders and pure depressions. In fact, cases of pure mania are quite rare. Most patients with mania experience depression in some other episode of the illness, and may even cycle back and forth from one pole of the illness to the other. Others experience only depressions.

Three separate family studies of mood disorders—one in Sweden (Perris 1966), one in Switzerland (Angst 1966), and one in the United States (Winokur and Clayton 1967)—established the now widely-used division between two types of mood illness, called unipolar and bipolar affective disorder. Bipolar affective disorders are those which include both kinds of episodes, mania and depression. Purely manic disorders are usually included in this category as well, even though strictly speaking they are not 'bipolar'. Unipolar affective disorders involve depressive episodes but never mania.

Later in this chapter we will discuss the genetic evidence for the unipolar–bipolar dichotomy along with theories of genetic heterogeneity in the mood disorders. For now, it is important that the reader be alert to the fact that some studies are based on samples of affectively disordered patients in general, while others are based on samples of patients with specifically either bipolar or unipolar illnesses. The unipolar–bipolar distinction may be important in genetic counselling since studies of unipolar illness may be irrelevant to families with a history of bipolar illness, and *vice versa.*

Other distinctions in mood disorder

Similarly, some studies are concerned only with severe forms of mood disorder (mania and depression) while others include non-psychotic affective disorders such as hypomania, cyclothymia (a personality pattern characterized by chronic, mild mood fluctuations), and chronic depressive disorders (mild depressions of long standing). Some types of data available for affective psychoses may not be available for the milder affective disorders, since the latter have not been studied as extensively as the more severe psychoses.

79

Mood disorders: mania and depression

Finally, primary affective disorders are sometimes distinguished from secondary disorders. Primary disorders are those which occur apparently of themselves, while secondary disorders, by definition, are those which occur in the presence of an already established psychiatric or major physical disorder. Mood disorders frequently occur as forerunners, concomitants, or complications of illnesses such as alcoholism, Huntington's disease, drug psychoses, and brain injuries. The genetic counsellor must establish through careful diagnosis whether the mood disorder in question stands by itself, whether it arises from a pre-existing illness, or whether perhaps the affected individual has two independent disorders simultaneously, as sometimes happens.

How common are mood disorders?

It can safely be said that mood disorders are some of the most common major psychiatric illnesses in modern society. Depression, in particular, is widespread, judging from the rate at which one encounters it or hears of it in everyday life. Accurate estimates of its prevalence, however, are hard to find. Such estimates should be based on studies which have used specific diagnositc criteria, personal interviews with a large random sample of subjects derived from well-defined catchment areas, and techniques for eliminating interviewer bias. But very few studies meet all of these requirements.

Some of the best available estimates have come from a well-designed community survey (Weissman and Myers 1978a, 1978b) carried out in New Haven, Connecticut, from 1967 to 1976, covering a large representative cross-section of the community.

This study found that in one year (1975–6), 4.3 per cent of those surveyed had experienced major depressions and 2.5 per cent had minor depressions, for a total of 6.8 per cent (5.7 per cent when only definite cases were counted). Nearly 1 out of 5 subjects over 18 years of age showed substantial depressive symptoms at that time.

If these figures seem surprisingly high, the lifetime prevalence figures are more surprising still. In all, 26.7 per cent, or more than 1 out of 4, reported major or minor depressions at some time in their lives (24.7 per cent when only definite cases were counted). The lifetime prevalence figures were far lower for mania (0.8 per cent) and bipolar affective disorder (1.2 per cent).

Were all of those who reported mood disorders seen by psychiatrists? Far from it. Of those with depressive symptoms, only 18 per cent said that they had been seen by a mental health professional. Another 80 per cent had brought their symptoms to the attention of a general physician, and a large proportion had not received any treatment.

Are mood disorders inherited?

Evidence from twin studies

Much of the most convincing evidence for a hereditary component in mood disorders has come from twin studies, although in the past couple of years a few adoption studies have also been reported.

Twin studies report widely differing twin concordance rates for mood disorders, probably because of differences in sampling techniques and diagnostic criteria, but possibly also because of cultural differences influencing the way mood disorders are expressed in different countries and climates. Researchers have reported identical-twin concordance rates as low as 50 per cent and as high as 93 per cent. Concordance rates for fraternal-twin pairs have ranged from 0 per cent to 39 per cent.

Yet even with so much divergence in the estimates, a pattern highly suggestive of genetic influence can be seen to emerge. It comes into clearer focus when the twin concordance data from seven major studies in the USA, Germany, England, Denmark, and Norway are pooled together. The identical- and fraternal-twin concordance rates from these studies combined are 76 per cent and 19 per cent, respectively. This is a highly significant statistical difference and persuasive evidence that mood disorder is a genetically inheritable disease—*if* it is safe to assume that no differences in home environment between identical-twin and fraternal-twin partners can account for the difference in concordance rates. If identical-twin partners tend to be treated identically while fraternal-twin partners are usually treated more as individuals, some of the higher concordance rate among identical twins might be accounted for by environmental factors.

A suitable test for that theory would be to examine identical twins raised separately from their partners under the hypothesis that concordance rates for mood disorder would be much lower when the element of shared family environment is cancelled out. Price (1968) examined the psychiatric literature for the scarce reports of affectively ill identical twins reared apart from their co-twins. He found 12 pairs, of which 8, or 67 per cent were concordant for manic-depressive disorder. Thus, identical twins brought up separately from their twin partners show intra-pair concordance for mood disorder at approximately the same rate as pairs brought up together. This result strengthens the evidence for genetic factors in mood disorder.

But as you already may have noticed, even identical twins, with identical sets of genes, do not inherit identical mood disorders as consistently as they inherit identical hair colour, body build, and other physical traits. Almost one-third of all the identical-twin pairs in the twin studies are discordant for mood disorders. That sizeable margin must mean that non-genetic (i.e. environmental) influences are interwoven with genetic effects in at least some cases.

Genetic evidence from adoption studies

Until very recently, no adoption studies of manic-depressive disorders had been reported in the psychiatric literature, even though adoption studies of schizophrenia had been underway in the United States and Denmark since the early 1960s. But in 1977 Drs Julien Mendlewicz and John D. Rainer reported study of bipolar mood disorder in the parents of adopted children conducted in Belgium.

The study focused on the biological and adoptive parents of a group of adopted children who had developed bipolar illness. This strategy would show whether or not the rate of mood disorder in parents was dependent on their being genetically related to an ill adopted child. The study included several groups of controls, including the biological and adoptive parents of normal adopted children and the parents of bipolar patients who had not been adopted away. Each of the parents was personally interviewed by a clinician who was kept uninformed of the relationship between the person interviewed and anyone else in the study. This precaution was intended to keep subtle biases of the observer from influencing the diagnostic ratings.

The most important finding of the study was that parents genetically related to bipolar adopted children had a significantly higher rate of psychiatric illness (especially mood disorders) than the parents who adopted and raised them (40 per cent vs. 16 per cent). Moreover, the rate of mood disorder in the biological parents was not statistically different from the rate of mood disorder in the parents of bipolar non-adoptees (31 per cent vs. 26 per cent). Thus, it appeared to matter little whether or not the biological parents were continually in contact with affected children as they grew up: both groups of parents developed mood disorders at approximately the same rate. In contrast, the rate of mood disorder in the genetically unrelated adoptive parents of bipolar subjects did not distinguish them from either the adoptive or the biological parents of normal adopted children. The researchers concluded that these results show the importance of genetic factors in causing mood disorder.

Further support for a strong genetic component in mood disorders has come from an American study (Cadoret 1978) which reported a significantly increased rate of depression in adopted children whose biological parents suffered from mood disorders.

How are mood disorders transmitted?

Knowing from twin and adoption studies that mood disorder is an inheritable illness, psychiatric geneticists have to account for the fact that it is not inherited in simple Mendelian patterns—the kind seen in pea plants and fruit flies. Drawing on the available data, they have postulated the following as possible models for the mode of transmission of mood disorder: X-linked dominance, autosomal dominance with reduced penetrance, or polygenes with quasi-continuous variation. None of these theories completely accounts for the hereditary patterns seen in mood disorder. They will be discussed below, but first we must consider the evidence that mood disorders might be transmitted by several distinct modes of inheritance.

Genetic heterogeneity

The subject of genetic heterogeneity in the mood disorders is full of unresolved questions too complicated to sort out here. The search for evidence would take us into the complicated field of psychopharmacological studies and biological studies of neurohormones, neuroenzymes, and monoamine metabolites. As others (Gershon *et al.* 1977) have admirably summarized these studies elsewhere, we will content ourselves with a short summary of the main hypotheses.

Since the time when the unipolar–bipolar distinction was first proposed (Leonhard 1957; Leonhard *et al.* 1962), information from family studies has suggested that unipolar and bipolar illness might be genetically separate illnesses. Family studies consistently show that relatives of persons with bipolar illness have a higher rate of mood disorder than relatives of persons with unipolar illness. Bipolar patients with depression consistently respond better to treatment with lithium carbonate than unipolar patients. Furthermore, twin study data (Zerbin-Rudin 1969; Perris 1974) on twin pairs who were concordant for mood disorder showed that 81 per cent of the pairs were also concordant for subtype (unipolar or bipolar). All of these observations strengthen the possibility that unipolar and bipolar are genetically distinct disorders.

Yet several other observations cloud the picture. In the families of bipolar patients, many researchers have found not only increased bipolar illness but also increased unipolar illness; in several studies, unipolar illness appears even more frequently in these relatives than bipolar illness (Angst 1966; Helzer and Winokur 1967; Goetzl *et al.* 1974; Gershon *et al.* 1975; James and Chapman 1975). These findings point to some overlap between unipolar and bipolar illness. The overlap is also seen in studies of drug response in mood-disordered patients. While bipolar patients respond better than unipolar patients to lithium used to treat acute depression, both bipolar and unipolar patients respond well to lithium used to prevent the recurrence of depression (Prien *et al.* 1973; Fieve *et al.* 1976; Coppen *et al.* 1976). This and other evidence from biological studies, conclude Gershon and his colleagues, suggest that if there is an underlying genetic division in mood disorder, it may not correspond well with the unipolar–bipolar distinction. The unipolar group, in particular, may include illnesses which appear to be closer to bipolar than to unipolar disorder.

These authors suggest two genetic hypotheses for the observed overlap between unipolar and bipolar illness. There may be only one genetic type, which appears as bipolar mood disorder when the genetic and environmental load is greatest and as unipolar disorder when the load is moderate. Or there may be two distinct genetic types; bipolar disorder would represent one type, but unipolar disorder would swing two ways, some cases being genetically related to bipolar disorder and others representing a separate genetic type. According to the second scheme, unipolar patients with a family history of mania may be genetically related to bipolar disorder. The question of a genetic distinction between unipolar and bipolar mood disorders cannot be resolved with the data available at present; more investigations will be needed.

Although the unipolar–bipolar division is not completely resolved, other attempts have been made to subdivide both unipolar and bipolar subtypes.

Winokur *et al.* (1971) studied age of onset in unipolar patients in connection with other variables such as family history of psychiatric illness and sex. These variables, they suggested, distinguish between two types of unipolar illness: an early onset (onset before age 40) called 'depressive spectrum disease' and a late onset type (onset after 40) called 'pure depressive disease'. The former is possibly a continuum of diseases related to depression, including alcoholism and antisocial personality. According to the authors, the prototype of depressive spectrum disorder would be a woman who becomes depressive before age 40, whose female relatives have a higher rate of depression than her male relatives, but whose male relatives seem to 'compensate' for their lack of depression with an increase in alcoholism and antisocial behaviour. Pure depressive disease, on the other hand, would be represented by a man who becomes depressive after age 40, whose relatives show a lower rate of depression, no difference between male and female relatives in rate of depression, and no large amount of alcoholism and antisocial behaviour among the males.

For reasons that are not entirely clear, British studies have not supported this division of unipolar mood disorder. The whole question demands further research, especially the fascinating suggestion that there may be a link of some kind between early onset depression and alcoholism or antisocial behaviour.

Bipolar mood disorder, too, according to Dunner *et al.* (1976), may possibly be subdivided into two genetically distinct groups, called BP I and BP II. The former combines episodes of depression with mania, while the latter shows depression altering with hypomania. BP II patients appear to have a higher rate of suicide, a later age of onset, and a different response to certain drug treatments than BP I patients. This distinction also has not yet been sufficiently established by research.

The spectrum of mood disorders

A spectrum of psychiatric illnesses, analogous to the spectrum of colours formed when light passes through a prism, is a continuum of disorders presumed to be somehow related in cause. Family studies have found that a number of disorders can be found with unusual frequency in close relatives of patients with mood disorder, thus suggesting that there may be a 'spectrum' of mood disorders.

Two kinds of disorder, minor depressive episodes and cyclothymic personality disorders match the statistical predictions of genetic models well enough that they may be considered part of a mood spectrum (Gershon *et al.* 1975). A certain proportion of schizoaffective disorders, those which show a family history and prominent symptoms of mood (affective) disorder, may be genetically related to mood disorders. Some studies have observed that alcoholism and anti-social behaviour occur at increased rates in families of people with mood disorders, but it is not yet clear to what extent genetic processes are at work in these disorders and to what extent their appearance among some relatives of persons with mood disorders may be influenced by cultural factors. Relatives of depressive patients have a higher rate of suicide than members of the general population, although the genetic connection between suicidal behaviour and mood disorder is unknown.

Models for the inheritance of mood disorders

Several genetic models can be dismissed at once as possibilities for the transmission of mood disorder because they do not fit family study data. Mood disorder is clearly not transmitted by a recessive gene placed on the X chromosome. If it were, the number of males with mood disorder would be expected to outnumber the females. That is because women who received the gene would usually be unaffected carriers, having two X chromsomes, one of which would probably contain an allele to overrule the recessive gene. In fact, however, all studies show that far more women than men develop mood disorders.

An autosomal recessive gene can also be ruled out. Since the parents of an ill person would most likely be unaffected carriers under this model, we could expect parents of manic-depressive patients to be less often affected than brothers and sisters. In fact, however, all immediate blood relatives are affected at approximately the same rate.

Autosomal dominance with reduced penetrance

Mood disorder is apparently not transmitted by an autosomal dominant gene that manifests itself in every case, since if its penetrance were complete (100 per cent), it would produce 100 per cent concordance rates in identical twins and would affect 50 per cent of all children, siblings, and parents of those with mood disorder. Actually, the rate of mood disorders in family members is closer to 15 per cent; to account for this rate, the single-gene dominant model would have to assume that the damaging gene finds expression in only about 30 per cent of the cases.

X-linked dominant genes in bipolar mood disorder

As early as 1938 (Slater 1938) it was suggested that bipolar mood disorder might be transmitted by a dominant gene located on the X chromosome. There is a relatively easy way of testing this theory: simply observe whether family studies turn up any pairs of fathers and sons who both have bipolar illness. If the hypothesis is correct, there should be none. The reason behind the hypothesis is that sons can only inherit a Y sex chromosome from their fathers. Hence, a damaging gene located on the X chromosome cannot be transmitted from father to son.

In a study of families of 89 manic subjects, Winokur (1970) investigated ill parent and ill child pairs. Strikingly, he found no ill father and ill son pairs, while the other three combinations (father–daughter, mother–daughter, mother–son) were each represented by at least 13 ill pairs. Such evidence more than hinted at an X-linked dominant mode of transmission for mania. Later studies, such as the adoption study of bipolar patients by Mendlewicz and Rainer (1977) have also found no ill father and ill son pairs.

Another test of the X-linked theory for bipolar illness capitalizes on the fact that gene loci for many specific traits have already been aproximately located on the X chromosome. Two of the X-linked traits are especially useful as positive markers on the X chromo-some: (1) the locus for red–green colour-blindness (which occurs in about 8 per cent of all white males) is found on the long arm of the X chromosome, and (2) the locus for the blood group labelled Xg^a is found on short arm. Other genes located near these sites will tend to be inherited together with them—they are said to be linked. Genes lying further apart on the chromosome will more often be separated by chromosome 'cross-over' during the formation of egg and sperm cells by meiosis and thus they will more often be inherited separately. Therefore, researchers regard evidence of linkage with a known X-linked trait as confirmation that the gene for the trait in question lies on the X chromosome.

Several studies using this strategy found significant evidence of linkage between mood disorder and colour-blindness loci as well as suggestive evidence of linkage with the Xg^a blood group (Winokur and Tanna 1969; Mendlewicz *et al.* 1972; Fieve *et al.* 1973).

The issue of X-linkage in mood disorder is not by any means settled. Numerous studies have found examples of ill father and ill son pairs. Furthermore, Gershon *et al.* (1977) have raised strong arguments against the probability of mood disorder being linked to both colour-blindness and the Xg^a blood group at the same time, citing the wide separation of their respective loci on the X chromosome.

It is possible that some of the son's illnesses in ill father and ill son pairs were actually transmitted from the mother's side of the family, although some of the studies checked against this possibility and did not find mood illness in the mothers. Another possibility is that some cases of mood disorder are X-linked while others have a different mode of transmission.

Information on X-linkage is not very useful for genetic counsell- ing at present, since it is rarely possible to determine by family history, and impossible to determine by clinical tests, which families have an X-linked form of the illness. Probably a great deal more must be learned about the positions of gene loci on the X chromo- some and the biochemical effects underlying mood disorder before the problem can be solved.

Given the present limitations on our knowledge, psychiatric genetic counsellors should not assume that bipolar mood illness is an X-linked disorder. Assuming X-linkage may lead to false esti- mates of risk and possible damage to the welfare of the clients.

Polygenes with quasi-continuous variation

Four features of mood disorder match the predictable character- istics of polygenically transmitted traits. First, cases of mood illness among relatives tend to be evenly distributed between the father's and mother's sides of the family. Secondly, the time of onset and course of the illnesses appear to be influenced by environmental variables. The fact that as many as one out of three manic-depres- sive identical twins have unaffected partners adds strong evidence that environmental factors may make a significant contribution. Thirdly, the risk of mood disorder to a brother or sister of an affected person increases with the number of other affected siblings. In polygenic inheritance, the number of affected siblings may show how strong a 'dose' of polygenes they have inherited from the parents (i.e. whether their liability is close to the threshold of disease or well above it). For classic single-gene traits, by way of contrast, the risk to all brothers and sisters remains the same regardless of the number already affected. Finally, the relatives of more severely ill manic depressives have a higher risk of affective disorder than relatives of those with milder cases.

A statistical technique called 'threshold model analysis' has been applied to family study data to determine which factors might best define thresholds of liability for manic depression and its subtypes. Another technique called 'segregation analysis' has been used to determine whether a single gene or multiple genes plus environmental influence best accounts for the pattern of inheritance seen in families. However, neither technique has yet provided conclusive results.

Inheritance of schizoaffective disorder

Some people have many of the major symptoms of schizophrenia but also experience major symptoms of mood disorder. In the past, these hybrid disorders have usually been thought of as special types of schizophrenia. Thus, they have been called by such names as 'atypical schizophrenia' and 'schizoaffective disorder'. But from twin and adoption studies discussed in this and the previous chapters, we have evidence that both schizophrenia and mood disorder are inherited illnesses and that they are genetically distinct from each other.

Is schizoaffective disorder, then, a subtype of schizophrenia? Or could it be a subtype of mood disorder? Or is it perhaps a third distinct illness bearing only a surface resemblance to schizophrenia and mood disorder? Or could it be 'all of the above'?

Many studies have suggested that there is a genetic relationship between mood disorder and at least some forms of schizoaffective disorder. Clayton *et al.* (1968) found that relatives of schizoaffective patients in their study had a higher rate of mood disorder than of schizophrenia. A family study of 'good prognosis' and 'poor prognosis' schizophrenics (McCabe *et al.* 1971) found that the good prognosis group (whose disorders were essentially the same by definition as schizoaffective disorder) had an elevated rate of mood disorder among their relatives while the poor prognosis group showed a higher rate of schizophrenia. A study of male MZ twins (Cohen *et al.* 1972) reported significantly higher twin concordance rates for schizoaffective disorder than for schizophrenia but found no significant difference between the concordance rates for schizoaffective and mood disorders.

Tsuang (1969) investigated 35 pairs of siblings, each of whom was diagnosed as having either schizophrenia, schizoaffective disorder, or mood disorder. When their ages of onset were analysed, schizoaffective patients more closely resembled the mood disorder group than the schizophrenia group, although a small group of schizoaffective patients appeared to be closely related to schizophrenia. In a further analysis of this same sample of patients (Tsuang 1979), the author tested whether schizoaffective disorder is a subtype of schizophrenia or mood disorder or whether it is a genetically distinct disorder by calculating expected numbers of ill siblings under each hypothesis and comparing these against the numbers actually observed. This analysis could not support the hypothesis that schizoaffective disorder is genetically distinct, but it did match both of the other hypotheses.

When all of these studies are compared, they lead to the conclusion that schizoaffective disorder can be separated into three subtypes, one in which mood symptoms predominate, another in which schizophrenic symptoms predominate, and a third in which both kinds of symptoms are marked or in which there is too little information to assign cases to either of the first two subtypes.

The psychiatric genetic counsellor may decide how to regard cases of schizoaffective disorder by referring to age of onset and family history information (Tsuang 1967). It may be treated as a mood disorder under either one of two conditions: (1) the age of onset is after 40 and the affected person has a family history of mood disorder, or (2) the age of onset is between 30 and 40 years and the affected person has a family history of schizoaffective disorder. When the age of onset is before 30 and there is family history of schizophrenia, schizoaffective disorder should be treated as schizophrenia.

What is the risk of inheriting mood disorder?

Family studies generally show that close relatives of mood-disordered patients develop mood disorder at about ten times the

What is the risk of inheriting mood disorder?

general population rate. Exact risk estimates are not available since the mode of transmission is not yet known.

Table 2 displays a summary of nine major family studies of mood disorder which were well-designed by current standards. The risks of mood disorder are broken down into separate rates of unipolar and bipolar illness in close blood relatives (parents, siblings, and children) of persons affected with unipolar or bipolar illness. The table also shows the total rate of mood disorder among the relatives. The risk figures provided by these family studies cover an uncomfortably wide range from lowest to highest reasonable estimates, largely because they differ among themselves in their methods of choosing samples and collecting data and in the criteria they used for diagnosing illness in relatives.

TABLE 2. *Risks of mood disorder in close relatives of affected persons**

Patients		Ill relatives	
		Range of estimates (low–high)	
	Unipolar (%)	Bipolar (%)	Total (unipolar + bipolar) (%)
Unipolar	7–19	0.3–2	8–23
Bipolar	6–28	4–18	11–42
		Males (%)	Females (%)
General population**		1.8	2.5

*Based on reivew of nine major studies in D. W. K. Kay (1978). Some data excluded.
**Based on population study in Iceland (Helgason 1964).

In the general population, as seen in the table, the rate of mood disorder is 2.5 per cent for women and 1.8 per cent for men, or roughly 1 out of 50 overall. Among relatives of unipolar and bipolar patients, the risks for all mood disorders combined jumps to between 10 and 20 times the general population rate. The highest risk shown here is for relatives of bipolar patients, up to 42 per cent of whom may also become ill. The lowest rate of disorder is the risk of bipolar illness among the relatives of unipolar patients (between 0.3 and 2 per cent).

The table supports the generally made observation that the relatives of bipolar patients have a higher rate of mood illness than the relatives of unipolar patients. Interestingly, the rate of unipolar illness is somewhat higher among relatives of bipolar patients than among relatives of unipolar patients—slightly higher even than the rate of bipolar illness among relatives of bipolar patients.

Very recently, a large follow-up and family study of mood disorders and schizophrenia conducted in the state of Iowa found the risk of mood disorder to be 17.8 per cent among close relatives of bipolar patients and 18.2 per cent among close relatives of unipolar patients (Tsuang, unpublished data 1979). These figures establish firm conservative estimates, in view of the fact that the study employed every available means of assuring reliability including the use of strict diagnostic criteria, a thorough mechanism for tracing and locating patients and their relatives, a large study sample with matched control groups, a 40-year follow-up period, blind personal interviews with nearly 80 per cent of the living study subjects and their relatives using an objective, structured interview form carefully tested for reliability, and elaborate computerized procedures for handling and analysing the collected data.

Table 3 shows how the risk to brothers and sisters of a person with mood disorder rises according to the amount of mood illness in the parents.

TABLE 3. *Risks for mood disorder in siblings, according to mood disorder in parents**

	Siblings affected (%)
Neither parent ill	12 (± 1.4)
One parent ill	26 (± 3)
Both parents ill	43 (± 19)

*Based on Winokur and Clayton (1967).

When neither parent of the affected person is ill, the risk to siblings (12 per cent) corresponds to the overall rates of mood illness in close relatives seen in Table 2. When one parent is affected, the rate more than doubles to 26 per cent, and when both parents are affected, the rate leaps again to 43 per cent.

Table 4 shows risks of mood disorder among immediate blood relatives of persons affected with mania. The figures are broken down according to sex of the relative and type of relationship. In each category of relationship (parent, sibling, children), over half of the female relatives are shown to be affected, and more than twice as many females are affected as males. Parents, siblings, and children are affected at roughly the same overall rate.

The figures in Table 4 may appear to be very high, especially the 83 per cent risk for daughters of manic patients. Several factors may account for the increase. First, the table lists risk estimates only for relatives of persons affected with mania, a form of mood disorder often characterized by earlier onset, greater severity of symptoms, and higher rate of illness among relatives than are usually seen in patients with unipolar affective disorder. Secondly, the figures in the table are based on a large study in St Louis which collected information on relatives not only from hospital records but also from follow-up studies and personal interviews with all available patients and their relatives. This rigorous method of research tends to disclose higher rates of illness in relatives than studies based only on data gathered from legal and medical records.

TABLE 4. *Risks of mood disorder in family members of manic patients**

Family relation	Risk of mood disorder (%)	
Mothers	55	(±8.7)
Fathers	17	(±8.8)
Sisters	52	(±8.7)
Brothers	29	(±9.3)
Daughters	83	(±15.3)
Sons	17	(±15.3)
All parents	41	(±6.9)
All siblings	42	(±6.6)
All children	50	(±14.4)
All female family members	56	(±5.8)
All male family members	23	(±5.8)

*Based on Winokur, Clayton, and Reich (1969); data are based on personal interviews and family studies of persons with bipolar disorder only.

Adapting risk estimates to individual cases

In adapting the categorical type of information found in tables of risk figures to his own case, the client may ultimately have to rely on hunches and educated guesses. His best guide will be a clear understanding of the factors typically associated with increased risk of mood disorder.

Mood disorders: mania and depression

The risk for all mood disorders is greater among women than among men, both for relatives of mood-disordered patients (see Table 4) and in the general population (see Table 2). Some studies suggest that female relatives of depressed patients may have twice the risk of mood disorder as male relatives, and female relatives of bipolar patients may have more than twice the risk of mood disorders as male relatives.

Age is another informative factor. When the affected family member has had an early onset of illness, say between ages 20 and 25, the risk of illness among that person's relatives tends to be greater. Relatives of persons with bipolar illness, if they have not yet reached middle age themselves, are at a greater risk for both bipolar and unipolar illness than those past that period of life.

People closely related to someone with bipolar illness have a greater risk for all mood disorders than those related to persons with unipolar illness. Risks increase among relatives of affected persons when there are many other affected relatives, especially when the relatives are within the immediate family circle. When one or both parents of the affected person are also affected, the risk of mood disorder for siblings greatly increases.

Conversely, the risk of mood disorder is lower among males, among those who are past the ages of greatest risk for bipolar illness, among relatives of persons with unipolar mood disorder (especially those with late onset depressions), and among those with few affected relatives. The risk of bipolar mood disorder among relatives of those with unipolar illness appears to be no greater than the general population risk for mood disorder.

This reminder is in order: some of the available risk figures for mood disorder are derived from studies using modern, strict criteria for unipolar and bipolar illness. Such figures would not strictly apply to relatives of persons with milder disorders such as cyclothymic personality disorder or hypomania. Yet milder disorders have to be considered in gauging risk, since a milder disorder may be mood disorder in a carrier state or the result of an incompletely expressed dominant gene, or possibly the result of polygenes operating below the threshold level of intensity. Where many such disorders are congregated along with a few causes of severe mood disorder in one family, they may indicate an increased risk of mood disorder to relatives within that family. Likewise, as the Winokur study of 'depression spectrum' disorders suggested, the presence of high

94

rates of alcoholism or antisocial behaviour in male relatives of a woman with early onset depression may point to an increased risk of mood disorder in that family. Finally, the occurrence of suicides within a family, even if for unknown motivations, points, though not conclusively, toward an increased risk of depression in relatives.

Lightening the load

The risk estimates for mood disorders are some of the highest among major psychiatric disorders, but they are not the whole story. The perceived burden of a mood disorder is at least as important a factor as risk in the psychiatric genetic counselling process. Some people may perceive the burden to be extremely great—those, for example, whose families have already been terribly disrupted because of mood disorders among their members. Such people may view the risk figures through a magnifying lens, and as a result, their fear may have a strong influence on their family planning.

Fortunately for some of those who are affected by or liable to mood disorder, the burden of the illness can sometimes be relieved in several ways. While some forms of mood disorder last for 30 years or more, they do not typically lead to gross deterioration in the victim's physical and mental health as chronic schizophrenia typically does. Between episodes of illness, the affected person may return to a normal or near-normal state of health. Moreover, some forms of mood disorder, especially depressions, occur relatively late in life when some patients are better able to bear any financial burdens incurred. Finally, because of great strides in psychophar-macology in recent decades, mood disorders have become some of the most successfully treatable of psychiatric illnesses. Electro-convulsive therapy (ECT), although it is sometimes viewed with suspicion by the general public, is still one of the most effective treatments for depression, offering measurable relief from symptoms without major side effects. Lithium carbonate is used effectively to treat mania and acute depressions in bipolar patients and is also useful in some cases as a preventive treatment against symptoms of depression. Other medicines called tricyclic anti-depressants are now used with great success in controlling symptoms of depression.

Since the discovery of effective drug treatments for mood disorders, the number of ill patients living in mental institutions has dramatically decreased. Many are now treated as out-patients while living at home in their communities. Paradoxically, such improvements in treatment may ultimately increase the incidence of mood disorders in the community by allowing more people with these illnesses than ever before to work, marry, and bear children. The need for genetic counselling for families with mood disorders may increase accordingly. At the same time, research on drug treatments in psychiatry has opened windows for researchers on previously hidden biochemical pathways in the body. These may help guide us someday to an understanding of the precise role of the genes in mood disorders. From such discoveries we may hope to learn of biological tests for the early detection of carriers of the damaging gene or genes and to develop new drugs to control or cure the disease.

Summary

Affective psychoses are severe mood disorders with two opposite manifestations, mania and depression, characterized, respectively, by extreme euphoria and extreme despondency. Strong evidence of a major genetic component in mood disorders has come from twin studies, supported by two recent adoption studies, although the exact mode of inheritance is not yet known. Data from twin and family studies fit various models of transmission, including X-linked dominance, autosomal dominance with reduced penetrance, or polygenic with quasi-continuous variation.

Several genetically distinct illnesses may be included within the mood disorders. Studies support a distinction between bipolar illness (episodes of both mania and depression) and unipolar illness (depression only, never mania). A possible genetic distinction exists between early and late onset forms of unipolar depression. Some cases of schizoaffective disorder may also be genetically related to mood disorder. Other disorders, including alcoholism, antisocial behaviour, chronic mild depression and a mild bipolar personality disorder may belong to a 'spectrum' of disorders genetically related to mood disorders. People with unipolar or bipolar mood disorders also have significantly increased rates of suicide.

The risks of mood disorder to relatives of affected individuals is roughly ten times higher than the general population rate. Family studies indicate that risks are considerably higher among women

Summary

than among men, among close relatives than among distant relatives of affected individuals, and among relatives of bipolar patients than among relatives of unipolar patients. The risk to brothers and sisters of an affected person rises steeply when one or both parents are also affected. Among relatives of unipolar patients the risk of bipolar illness appears to be very low.

The burden of mood disorder is lessened by the fact that episodes of illness may be separated by periods of normal functioning, the fact that many forms of mood disorder do not occur until relatively late in life, and the fact that modern treatments are sometimes quite effective in reducing major symptoms of mania and depression.

6

Presenile dementias

Dementia is a loss of previously attained intellectual abilities severe enough to disrupt a person's family, professional or social life, and where the loss is due to progressive physical deterioration of the brain, apparently an acceleration of a natural process of ageing. Dementia is most widely known as a disease of old age but can also occur in middle age.

In this chapter we will discuss dementias that occur before old age, since their time of onset, their severity, and their tendency to run in families make them the type of mental illness for which genetic counselling is often necessary.

The discussion first turns to Huntington's disease, a hereditary dementia usually mixed with a severe movement disorder called chorea. Some have called it 'the most vicious disease known to man'. Then we consider the classic forms of dementia with onset in middle age: Alzheimer's and Pick's diseases. These two diseases are virtually indistinguishable on the basis of observable symptoms, although they can be differentiated by close study of affected brain tissue. Because the differences between the two are beyond the technical range of this book, they will be referred to collectively as 'presenile dementia'.

Symptoms of dementia

Dementia typically enters under disguise. It may be heralded by changes in personality easily mistaken for signs of exhaustion or anxiety. The victim may lose interest in a job, hobbies, or favourite sports. Some want to sleep all the time or complain about the energy needed to do ordinary tasks. Relatives and business associates may remark that the person is 'not himself lately'. Normally careful workers may forget simple things, get frustrated easily, turn in sloppy work, and neglect their personal appearance. New information becomes difficult for them to absorb. Some begin to use poor judgment and lose their ability to manage their own private affairs.

A former teetotaller may suddenly become a heavy drinker. For no apparent reason, a normally modest person may begin to use foul language or act seductively toward complete strangers. Some experience frequent unprovoked outbursts of anger.

Along with personality and behavioural changes, many victims of dementia develop psychotic symptoms including depressions, mania, delusions of grandeur or guilt, and hallucinations. Many are already under the care of a psychiatrist, sometimes diagnosed as being alcoholic or having schizophrenia or a mood disorder, before the underlying organic process of the disease can be discovered.

Many demented patients die of the illness about 10 to 20 years after the onset of symptoms, if suicide, accident, or infectious disease has not claimed their life earlier. Those who live out the full course of the illness may become bedridden and helpless years before the end of their life.

Huntington's disease

A quarter of a century before Mendel's laws of inheritance became known to the scientific and medical community, a young American doctor named George Huntington, writing in the *Medical and surgical reporter* ('On chorea', 13 April, 1872) accurately described the hereditary pattern of a form of presenile dementia which still bears his name:

> The hereditary chorea, as I shall call it, is confined to certain and fortunately a few families, and has been transmitted to them, an heirloom from generations away back in the dim past.
>
> When either or both the parents have shown manifestations of the disease, and more especially when these manifestations have been of a severe nature, one or more of the offspring almost invariably suffer from the disease, if they live to adult age. But if by any chance these children go through life without it, the thread is broken and the grandchildren and great-grandchildren of the original shakers may rest assured that they are free from the disease. . . . Unstable and whimsical as the disease may be in *other* aspects, in this it is firm, it never skips a generation to again manifest itself in another; once having yielded its claims, it never regains them.

Dr Huntington was accurately describing a hereditary pattern that would now be called an autosomal dominant mode of transmission. Each child of an affected parent, whether male or female, has a 50 per cent chance of inheriting the gene, and anyone who inherits the gene will sooner or later develop the disease. There are no unaffected carriers, as there would be in a recessive condition. If the disease skips one generation, future generations in that family will not be affected. (The exception to this rule is when an adult who possesses the damaging gene has children but dies of other causes before Huntington's disease appears. In such cases, the disease may seem to skip a generation, but the gene may be passed on to the children.)

It used to be thought that Huntington's disease would typically begin when the victim was in his or her middle thirties, but with better and more long-term family studies, we now know that the average age for developing Huntington's disease, is 44 years. On very rare occasions, it appears in children as young as 8 or in adults as old as 75.

Dr Huntington mentioned in passing the rarity of the disease, and in this, as in many of his other observations, he was correct. According to community surveys, Huntington's disease is found in barely more than 1 out of every 25 000 people in the general population. This, of course, is no consolation to people born into affected kinships, for their family trees are typically loaded with cases over many generations. Studies have shown, however, that the relatives of Huntington patients also have an increased risk of a variety of other problems such as suicide, convulsive disorders, mental deficiency, alcoholism, behavioural abnormalities, and criminality. The increase of these disorders in relatives may be partly due to un-detected cases of Huntington's disease in their early stages before the processes of dementia and chorea have become far advanced.

The burden of Huntington's disease

A crueller disease than Huntington's could hardly have been designed by a committee of devils. Most of its victims lose control of both their minds and their bodies in a long process of deterioration lasting anywhere from a few years to a few decades. It is highly hereditary, but because of its onset in middle age, many of its victims pass on the fatal gene to their children before they even suspect that they may have the disease themselves. Because it is so inheritable, it can cause not only its victims but also their children and grandchildren and all other close blood relatives to live their lives in a state of fearful suspense.

Huntington's disease

Huntington's disease differs from other dementias in the fact that it typically includes a movement disorder with the symptoms of dementia. Some first notice a change in their movements when their handwriting becomes illegible. They may try to cover up the change with excuses, jokes, or avoidance of situations that may require them to write. Some patients notice upsetting internal tremors and twitches beginning years before any outward changes appear. Eventually the disturbance begins to affect the way the person walks, giving him or her an odd, off-balance lurch which onlookers may mistake for a drunken swagger. Fingers and toes begin to twiddle uncontrollably, shoulders and sides hitch up like a marionette's, arms flail purposelessly in the air. People who are at risk for Huntington's disease sometimes begin to fear all signs of clumsiness in themselves, interpreting them as sure signs that they have inherited the dreaded disease.

For centuries observers have felt that the movements resembled a tragically disfigured dance. The movements thus came to be called 'chorea', from the Greek word for dance. After Huntington's vivid description of the disease, it came to be called Huntington's chorea, but the accepted medical name has recently been changed to Huntington's disease in recognition of the fact that not all victims of the disease develop a choreiform movement disorder. Some victims, particularly those who develop the disease at a young age, become progressively rigid without choreiform movements.

In the final stages of the disease, the chorea may lead to complete immobilization, and if the victim does not first die of other causes— accident, infectious disease, suicide—he or she may spend the last few years of life bed-ridden, completely dependent on nursing care.

Any mental illness makes victims not only of the affected person but of all those who are dependent on him or her for love, companionship, leadership, or material support. In no disease is this more true than in Huntington's disease. The impaired judgment of the affected person, the tendency of the disease to masquerade as another form of illness during its early phase, and the onset of illness in middle age create a network of suffering and worry in which all the members of an affected family can become enmeshed.

Case history

The following case history shows concretely how Huntington's disease places both a physical and a psychological burden on everyone in the family:

At age 46, the owner of a wholesale warehouse—call him William Taylor—was diagnosed as having Huntington's disease. Some of his telltale symptoms were slurring in his speech, difficulty in chewing and swallowing, clumsiness, and loss of memory. A thorough scouring of medical records disclosed the fact that Mr Taylor's mother and one aunt had both died of Huntington's chorea, while a great-uncle on his mother's side had lived his last eleven years isolated in a mental institution with what in retrospect probably should have been diagnosed as Huntington's disease. No one in Mr Taylor's family had ever spoken about these deaths. He remembered his mother only as a bedridden invalid, but he had been told that she died of multiple strokes.

The diagnosis was the climax of a long period of growing hardship for the Taylors and their four children. Eight years earlier Mr Taylor had begun to drink heavily, even though he had always before been a light drinker. A string of business mistakes perhaps related to his disease—placing redundant orders, failing to keep up his inventories, and angrily firing a trustworthy foreman—brought him near bankruptcy. He would make phone calls forgetting that he had already transacted that same business the previous day. He would come home from work irritable but unable to explain the reason for his anger. Sometimes he would shout obscenities at his wife and children over trivial disagreements, and a few times his anger spilled over into physical abuse. His outbursts led to visits to doctors, counsellors, and psychiatrists over the next several years. But before his illness could finally be diagnosed correctly, he had been treated for assumed manic depression, alcoholism, and atypical schizophrenia.

Mr Taylor finally lost the warehouse and tried without success to hold down part-time work. The cost of consulting doctors, buying medications, and missed work forced Mrs Taylor to find a job to support the two youngest children who remained at home. Quite irrationally in the midst of all this unsettlement, Mr Taylor often expressed his wish to have more children—another handful, as he phrased it—against his wife's firm objections.

When the presence of Huntington's disease was discovered and explained to the family, the distresses of the past eight years fell into better perspective. But the family received a new series of shocks when a genetic counsellor explained to them that the disease is transmitted by a dominant gene and that each of the Taylor children would therefore have a 50 per cent chance of developing the disease. This was crushing news to the family, especially to the two oldest sons, who understood the implications most clearly.

The oldest son Bill had already married and was now the father of two little girls. He understood that if he had received the damaging gene from his father, each of his daughters would also have a 50 per cent chance of developing the disease. Bill's wife was now pregnant with their third child. But upon receiving the news that Bill's father had Huntington's disease, their glad anticipation as expectant parents turned into apprehension. They were completely unready emotionally and intellectually to confront the thorny ethical questions surrounding the options of abortion, adoption, artificial insemination, or carrying the pregnancy through to full term.

The second son Dave had a different reaction. Engaged to be married when he was informed of his father's diagnosis, he soon afterward broke off the engagement. He appeared to accept his own risk of illness fatalistically, but inside he harboured feelings of worthlessness and resentment. He spoke of feeling 'marked for life, an untouchable'. He swore that he would never fall in love again, never marry, and never have children. Despite his mother's appeals for love and understanding, he was repelled by his father's illness and began to avoid other members of his family.

Mrs Taylor, though not genetically at risk herself, was caught in the middle of the emotional and financial turmoil. The responsibility for managing the household fell increasingly on her. Her husband would soon require the service of a professional nurse, but for the present she was responsible for those duties. She knew that she was watching an unstoppable process of deterioration. Meanwhile, she had to help her children confront their own fear and grief. She could see painful uncertainty stretching twenty years and beyond in her family's future as each child and possibly some of the grandchildren would enter adulthood not knowing whether they had inherited the fatal gene.

As Mr Taylor's disease progresses, this family will need a great deal of professional care, and each member could benefit if informed, compassionate genetic counselling were to become a regular and integral part of it.

Genetic counselling for Huntington's disease

Genetic counselling for Huntington's disease *must* begin with a clear diagnosis of the disease. But in order for a proper diagnosis to be made, other diseases resembling Huntington's disease must first be ruled out. The presence of chorea in the patient is not enough evidence by itself to prove that the disease is Huntington's, since similar choreiform movements occur in many other disorders, including Parkinson's disease, multiple sclerosis, certain disorders of metabolism, and some infectious diseases. Some of these are not hereditary and some may be treated with proper medication or manipulation of the patient's diet.

A second essential ingredient in a proper diagnosis of Huntington's disease is a thorough family medical history. This may be a tedious process for the counsellor, involving letter-writing, telephoning, searching through records, and interviewing family members. The counsellor must investigate and record the name, age, sex, and health of each living family member, as well as taking note of adoptions, stillbirths, abortions, miscarriages, cases of doubtful paternity, or the existence of half-siblings. But most importantly, the counsellor must try to ascertain the psychiatric history of all family members, living and dead, for several generations. All of this information may be needed to detect the telltale pattern of autosomal dominant transmission. In large families, the pattern may be tragically obvious—families with nine or more children, as many as five of them affected by Huntington's disease, are not unheard of.

In a few cases, unfortunately, a clear diagnosis will be impossible to make, even after all the available data have been collected. The symptoms of the disease may not match the typical signs and symptoms of Huntington's disease, the family history may be impossible to complete, or the completed family history may yield ambiguous information. The family confined to this diagnostic limbo has a special kind of burden to bear: they must experience all of the fear of inheriting a disease, but they have none of the certainty that the disease is in fact inheritable and no clear idea of the risks to various members of the family. Unfortunately, the counsellor's hands are tied until either the passage of time produces new information pertaining to the diagnosis (e.g. a change in the patient's symptoms or new family history information) or medical science devises a reliable clinical test to establish the presence of the disease. The genetic counsellor, meanwhile, owes such families a full measure of humane support.

Once the diagnosis of Huntington's disease has been communicated to the patient and any involved family members, once the counsellor has helped them cope with emotional conflicts almost invariably stirred up in family members by such a diagnosis, and once the counsellor has assessed their individual needs as well as their levels of emotional and intellectual maturity, the subject of risk is ready to be considered.

It is easy to estimate the likelihood that a child will inherit the damaging gene for Huntington's disease from an affected parent: the risk is 50 per cent for both sons and daughters. But when a person enters the standard period of risk for developing Huntington's disease (roughly the years between ages 20 and 55) without becoming affected, the person naturally would like to know the likelihood that he or she is among the fortunate 50 per cent of at-risk individuals who do not inherit the gene. Present good health is, of course, a favourable sign, and the longer the person survives without becoming ill, the more likely it is on a purely statistical basis that he or she has not inherited the damaging gene. However, the at-risk person may have to live with the fear of inheriting the disease for a whole lifetime, since the risk is never completely eliminated.

Because Huntington's disease often waits until the fourth or fifth decade of life to appear, it sometimes happens that an affected person has an adult son or daughter who is, as yet, unaffected but who, in turn, has borne a child. Unaffected parents caught in the middle of this situation typically want answers to the questions: 'What is my risk of inheriting the disease?' and 'What is the risk to my child?'

These questions can be answered with data based on our knowledge of the exact mode of transmission of Huntington's disease combined with data based on studies of age on onset in patients. The table offered here shows risk estimates for unaffected sons of patients and for the children born to those sons. The figures are modified downward according to the unaffected son's age. For example, if the son of a man with Huntington's disease reaches age 20 without developing the disease, his age-adjusted risk, as seen in the middle column, is reduced from 50 per cent to 44 per cent; his young child—the grandchild of the man with Huntington's disease—now has an age-adjusted risk of 22 per cent. If the young man survives until 49 years of age without developing the disease, his risk will fall to 9 per cent and his child's, again, will be half that, or 4.5 per cent. Because Huntington's disease affects men and women in equal numbers, the figures in the table ought to be approximately accurate for women as well as men.

For readers who are curious to know the basic arithmetic used to compute age-adjusted estimates like those in the table, the following illustration may be helpful. If we choose as our sample a young man of age 29, the problem is to explain how the 40 per cent risk figure listed in the table could be obtained. The starting point for such calculations is data concerning the age-of-onset pattern in Huntington's disease. For our sample we need to begin with the fact that, according to age-of-onset studies, 1 out of 3 victims of Huntington's disease are already affected by age 29, while the remaining 2 out of 3 become affected later in life. Of course, if we wished to calculate all of the other figures in the table, we would need to consult the entire age-of-onset curve established for Huntington's disease.

Let us imagine a Mr X whose father has Huntington's disease but who himself has reached age 29 without yet becoming affected, even though the possibility remains that he might have inherited the damaging gene from his father. At birth, Mr X had a 50 per cent lifetime chance of getting the illness (and, likewise, a 50 per cent chance of not getting it). But now as he enters further into adulthood, he wants to know how much his risk is lowered by the fact that he has passed part of the way through the risk period without becoming ill.

TABLE 5. *Risk of Huntington's disease to a young child whose father is normal* but whose grandfather is affected***

Age of father	Father's age–adjusted risk (%)	Risk for young child (%)
20–24	44	22
25–29	40	20
30–34	33	16.5
35–39	25	12.25
40–44	14	7
45–49	9	4.5

*Father's risk, unadjusted for age, is 50 per cent.
**Data from Nora and Frazer (1974)

Age of onset studies show, as mentioned above, that of all those who actually receive the damaging gene, one out of three become affected by age 29 and the other two out of three become affected later in life. For Mr X to be a still-unaffected possessor of the damaging gene at his age, he has to have met both of two conditions (each with its own probability): he has to have inherited the gene (the probability of this is 50 per cent) and, among all those who possess the gene, he has to belong to the group who develop the disease after age 29 (the probability of this is $2/3$). The chance of his meeting both conditions at once is expressed as the first probability multiplied by the second: $1/2 \times 2/3 = 2/6$ (or $1/3$). So we have found that Mr X has a ⅓ probability of being an unaffected possessor of the damaging gene now at age 29.

But we are not finished, for we have described only one out of two possible situations that might explain why Mr X is still not affected. He might have the gene and still be unaffected, or he may not have inherited the damaging gene at all. The probability of the first, as we have shown, is ⅓. The probability of the second is the basic 50–50 risk of inheriting a dominant gene, or ½. There would be no sense in multiplying these fractions together as we did above. That was the method used to find the probability of two conditions being met simultaneously, but here we have mutually exclusive options: either Mr X has not inherited the gene, or he has inherited it without yet being affected. Obviously, both cannot be true of one person at the same time. We must use a different method, as follows, to compute the probability that one condition will be met to the exclusion of others.

The entire group of persons at risk for Huntington's disease can be divided into three basic groups: those who did not inherit the damaging gene (and are therefore unaffected), those who inherited it but are not yet affected, and those who have inherited the gene and have already become affected. Mr X belongs to the subgroup which is unaffected. Thus, he wants to know, out of all those like himself—sons of Huntington's patients, age 29, not yet affected with the disease—what percentage actually carry the damaging gene? The answer will serve as an age-modified estimate of his estimate of his risk of developing Huntington's disease.

We can set about finding the answer by the following method. We have already said that Mr X's chance of not inheriting the damaging gene is ½, while his chance of having the gene without yet being affected is ⅓. Expressed using the lowest common denominator, these ratios translate into ³⁄₆ and ²⁄₆, respectively. Theoretically, then in a group of six 29-year-old sons of Huntington's patients, three should have the normal gene and be unaffected. These two groups together account for five of the six members; the one remaining member should have the damaging gene and be affected. The ratio of the three groups to each other is thus 3:2:1. (As predicted, half (2 + 1) receive the damaging gene and half (3) do not, while of those who do receive the damaging gene, one out of three is affected by age 29 and two out of three are not.)

Mr X's question was this: of all unaffected 29-year-old sons of

Huntington's patients like myself, what percentage actually carry the damaging gene? In our group of six, five were still unaffected. But out of those five, two possessed the damaging gene. A ratio of 2 to 5, expressed as a percentage, is 40 per cent. This, then is the answer to Mr X's question, as it appears in the table: Mr X, age 29, at risk for Huntington's disease but still unaffected, has a 40 per cent age-modified risk of carrying the gene for Huntington's disease. It must be remembered of course, that *if* he does carry the damaging gene, he will inevitably suffer the disease.

The risk to Mr X's child is easy to calculate, starting from this figure. For the child to become affected, Mr X must be affected (the probability of this is 40 per cent, as we have just seen), *and* the child must inherit the gene (the probability is 50 per cent). The product of the two probabilities is 20 per cent, as the table shows.

If Mr X grows older without developing the disease, his own risk must be continually revised downward in accordance with empirical data, and his child's risk must in turn be lowered. But note that if Mr X actually develops Huntington's disease in spite of his lowered statistical risk, all the figures in the table must change. They are based on his exact risk of 50 per cent for inheriting the damaging gene from his affected father. If he inherits the disease, his risk could be expressed as 100 per cent. The risks presently listed for him would thereafter apply to his child. On the other hand, if Mr X passes entirely through the risk period without developing the disease, it can quite safely be assumed that the gene has not been passed on from Mr X's father to him and that future generations in that family will not be affected.

The Committee to Combat Huntington's Disease (CCHD)

Risk estimates have an important role to play in genetic counselling for Huntington's disease, especially for people whose acquired notions of their inherited risk are wildly out of line. But statistical estimates too easily leave the impression that the struggle against Huntington's disease, or any other hereditary disorder, is fought on paper between platoons of numbers. The tendency to treat victims of Huntington's disease as non-persons and to exclude them from the human mainstream has always been too great—even among patients themselves. Genetic counsellors and other agents of professional care have the opportunity to oppose that type of injustice rather than to perpetuate it.

Huntington's disease

Fortunately for sufferers of Huntington's disease and their families, there exists a strong organization of volunteer and professional workers, the *Committee to Combat Huntington's Disease* (CCHD), dedicated to using every available resource to humanize the care, treatment, and counselling for Huntington's disease while encouraging basic biomedical research, trying to influence legislators on behalf of Huntington's victims, raising funds, and disseminating critically needed information to thousands of families and organizations all over the world.

The committee is a monument to the persistence of a dedicated lady, Marjorie Guthrie. Her late husband, the legendary American folksinger Woody Guthrie, died of Huntington's disease in 1967 after 15 years of struggle through misdiagnoses of his illness and confinement for two years in a psychiatric ward. Near the end of his life, Marjorie Guthrie was able to prove to doctors, despite their scepticism, that her husband's mind was still alert within his hopelessly incapacitated frame. Such experiences opened her eyes to the worldwide need for better research, treatment, and public information about Huntington's disease. She formed CCHD in 1967 to help bring these changes about. Her husband's fame, the popularity of her son Arlo Guthrie (composer of 'Alice's restaurant') and her own imaginative and relentless efforts soon raised the committee into the public spotlight.

Today, more than a decade later, the committee is a leading national voluntary health organization with a membership of more than 17 000 and with chapters and branch committees in 31 states in the United States. It maintains a communication and support network for about 7000 families with Huntington's disease listed in the committee's files. Through its many local representatives it presses the effort to reach the thousands of Huntington's families who have not yet been identified. Similar organizations have taken shape after the example of CCHD in the United States, Canada, United Kingdom, Belgium, Holland, and Australia.

CCHD, with Marjorie Guthrie now as its president emeritus, carries out its work among families, congressmen, medical researchers, doctors, and counsellors in a spirit that discourages gloom-spreading and resigned pity for Huntington's patients. Guthrie hates it, she has said, when Huntington's disease is described as fatal. She mocks that attitude: 'Fatal implies that your life is over tomorrow. Life itself is a fatal disease and being born is the first symptom.'

Instead, CCHD is working to encourage team approaches to counselling and treatment for Huntington's disease which would co-ordinate diagnostic services, medical supervision, nursing care, family genetic counselling, social and legal services, and so on, emphasizing the remaining capacities of the Huntington's patient for work, pleasure, and communication rather than defining the patient in terms of his or her disabilities.

Classic presenile dementias: Pick's disease and Alzheimer's disease

The two classic forms of presenile dementia have names few people know: Alzheimer's disease (named after the nineteenth-century German neurologist, Alois Alzheimer) and Pick's disease (named after the ninenteenth-century Czechoslovakian psychiatrist, Arnold Pick). Both of these diseases usually set in between the ages of 40 and 65; the mean age of onset for Alzheimer's disease is estimated at 44 years and for Pick's disease at 48 years.

For all practical purposes, Alzheimer's and Pick's can be grouped together under the name presenile dementia although the two cause deterioration in different parts of the brain and could be distinguished by a trained medical analyst studying damage to brain tissue under the microscope. Both cause symptoms typical of dementia: personality change, memory loss, decay of intellectual reasoning powers, disorientation, and mental symptoms resembling schizophrenia, mania, or depression. Unlike Huntington's disease, presenile dementias do not typically include a movement disorder along with the symptoms of dementia.

Shrinkage and destruction of brain tissues are clearly associated with presenile dementia, but no strict one-to-one relation can be drawn between the amount of cell deterioration and the severity of symptoms. Some severely affected patients show relatively little tissue damage on postmortem examination of the brain. Other patients with less severe symptoms have shown extensive degeneration of brain cells. Such degeneration is a marked acceleration of what appears to be a natural process of ageing. Studies have shown that virtually every one who reaches the age of 100, regardless of their state of health, has developed some brain tissue damage of the sort seen in demented patients, and many people develop such damage years earlier. Unfortunately, no one has yet satisfactorily explained what causes the deterioration. And once the process has begun, it cannot be stopped or reversed by any known treatment.

Genetic counselling for Alzheimer's and Pick's diseases

Both Alzheimer's and Pick's diseases show a strong tendency to run in families, thus suggesting that they may be genetically transmitted. Since these diseases, like Huntington's disease, usually occur in middle to late-middle age, many of their victims have started families long before they become aware that their children are genetically vulnerable to the illness. The shock of such a realization and the worry that may follow are proper matters for a family to deal with under the guidance of a genetic counsellor.

The critically important first step in genetic counselling for presenile dementias, as in all genetic counselling, is to verify the diagnosis of the affected person's illness. This is an especially important obligation in the dementias because typical symptoms of dementia can be produced by a number of disorders where brain-tissue degeneration is not necessarily present. Some of these look-alike disorders can be successfully treated. A short sampling of disorders besides Alzheimer's and Pick's diseases which can produce symptoms of dementia would include the following: chronic abuse of alcohol or drugs (bromides, barbiturates, phenothiazines, etc.), vascular diseases (e.g. atherosclerosis), normal pressure hydrocephalus, brain tumours, vitamin B_{12} deficiency, infectious diseases, abnormalities of the endocrine system such as hypothyroidism, chronic lung or kidney disease, and chronic hypoglycaemia. Establishing a diagnosis of Alzheimer's or Pick's diseases must necessarily be preceded by a long process of elimination.

The genetic counsellor must also complete a thorough survey of the affected person's family psychiatric history. This may help to clear up a problem of diagnosis, if one exists, or reveal the genetic mode of transmission in a few cases. Studies by Heston (1966, 1977) and Shenk (1959) found through close investigation of family histories that a small minority of familial cases of Alzheimer's and Pick's diseases follow an autosomal dominant pattern of inheritance.

In those rare kindreds where an autosomal dominant pattern is detectable, the exact risk of dementia to immediate blood relatives of an affected person is 50 per cent, and 25 per cent for those relatives one step further removed (nieces, nephews, uncles, aunts, and grandchildren).

111

Presenile dementias

In the great majority of cases, however, the immediate relatives have a much lower risk, as seen in Table 6.

TABLE 6. *Risks* of presenile dementia in first-degree relatives of affected individuals***

Related to affected person	Risk rate (%)
Parent	15.0
Brothers and sisters	
All	5.0
One parent affected	16.0
Neither parent affected	2.5

*Corrected for age.
**Based on Sjogren *et al.* (1952).

The table shows that about one out of six parents who have an affected child also have developed dementia. The rate among brothers and sisters of affected persons is only one-third as high overall as the parents' rate. However, the breakdown of figures for brothers and sisters shows that when one parent of an affected person also has dementia, the risk to brothers and sisters of the affected person is 16 per cent, compared to 2.5 per cent when neither parent is affected. This type of pattern suggests in most cases a polygenic mode of transmission, since the risk seems to rise and fall in accordance with the degree of genetic loading.

These risks are not high in comparison to those for Huntington's disease, but they are from 25 to 150 times the combined rate of Alzheimer's and Pick's diseases in the general population, estimated at 0.1 per cent.

Obviously, it is important for the genetic counsellor to gather family information very thoroughly and as accurately as possible in drawing up a pedigree, so as not to miss a pattern of autosomal dominant transmission in the kindred. Counselling an autosomal dominant family as if the disease were transmitted polygenically in their case would cause the relatives seriously to underestimate the risk that dementia might recur in their family.

Since the dementias have a higher age of onset and a lower

112

prevalence in the population than mental disorders such as schizophrenia, alcoholism, or manic depression, cases come less often to the attention of psychiatric genetic counsellors. But since dementias are mental illnesses chiefly of middle and old age, any process in civilization that increases the average life span for members of the general population will probably increase the rate of dementias as a proportion of all serious mental and physical diseases. Thus, if the threats at the frontiers of medical science—cancer, heart disease, arteriosclerosis, birth defects—are assaulted one by one and overcome, extending the normal life span in countries where modern medical technology is available, dementia may come more and more into the medical and psychiatric spotlight. At the same time, it can be expected that the need for genetic counselling for demented patients and their relatives will increase apace.

Summary

Dementias are inheritable disorders, usually occurring in middle, late-middle, or old age, characterized by personality changes, psychiatric symptoms, and the victim's loss of previously attained intellectual functions due to physical degeneration in the brain.

The two classical forms of dementia, Alzheimer's and Pick's diseases, appear to be polygenically transmitted, with risks to siblings of affected persons in the range of 5 per cent (16 per cent when one parent is also affected). A few families in both diseases show an autosomal dominant pattern of transmission, suggesting that the dementias may be genetically heterogeneous.

Huntington's disease, a central nervous system disorder which causes dementia often (but not always) mixed with a choreiform movement disorder, is known to be caused by an autosomal dominant gene with age-dependent penetrance. It occurs at a mean age of 44 years and usually causes death within 10 to 20 years after the first onset of symptoms. The risk to first-degree relatives of an affected person is high—50 per cent—but the empirical risk must be modified downward on the basis of age-of-onset data as the genetically vulnerable person passes further into the standard period of risk without developing the disease. Huntington's disease affects men and women in equal numbers and does not appear in a family once it has skipped a generation. *The Committee to Combat Huntington's Disease* (CCHD) has changed the face of counselling and treatment for Huntington's disease through its salutary efforts to support

Huntington's families, inform the general public, and encourage basic biomedical research.

Genetic counselling is especially necessary to families affected with dementias, since victims typically pass into the childbearing years before the disease is manifested. Unfortunately, there is no known treatment for true dementias other than therapy to prolong the victim's useful life, supportive arrangements in the home and institution, and medication to control treatable symptoms. Should average life spans increase as a result of breakthroughs in medical science, the overall incidence of dementia and consequently the need for genetic counselling for dementias are likely to increase.

7

Alcoholism

What are your chances of becoming an alcoholic? The life time rate of alcoholism in the general population is more than 5 per cent for men (less than 1 per cent for women), according to a family study in the midwestern United States (Winokur and Tsuang 1978). If you are a male and one of the members of your immediate family is an alcoholic, your chances of becoming an alcoholic sometime in your life may be as high as 50 per cent, as shown by a wide study of the relatives of alcoholics (Winokur *et al.* 1970).

The high rate of alcoholism in male relatives of alcoholics highlights the strong familial pattern in alcohol abuse. Studies of twins and adopted children show, moreover, that a strong genetic influence operates in severe forms of alcoholism. The hereditary nature of alcoholism qualifies it as a proper subject for genetic counselling. In this chapter, we discuss research findings of interest to those who are concerned about the possible recurrence of alcoholism in their family.

Case report

The following fictionalized account of an actual case of chronic alcohol dependence shows the type of burden that can be imposed by a longstanding alcoholic disorder:

Eugene Bishop, a 43-year-old white male, co-owner of a small real-estate firm, presented himself at a mental health centre because of complications related to a longstanding drinking habit. He had started to experience frequent blackouts; his hands and head were trembling; and he had recently been jailed for injuring a child on a bicycle while driving under the influence of alcohol. He had been hospitalized twice before to 'dry out' after benders lasting up to three weeks at a time. Each time he had presented himself at the emergency ward of a local hospital with withdrawal symptoms: palpitations of the heart, nausea, a crawling feeling on his skin, headaches, sweating, severe anxiety, and an oppressive suspicion that his business colleagues were conspiring to ruin him. Each visit was followed by numerous out-patient visits as he tried repeatedly but unsuccessfully to control his compulsion to drink.

The present complications stemmed from a time shortly after a Colorado vacation during which he had managed to limit his drinking to not more than two cocktails per day. However, he had returned to find conflicts among his subordinates at the real-estate office. His declining capacities as a supervisor because of his drinking habits were partly to blame for the tension; the co-owner of the firm knew of his problem and had renewed pressure on him to sell out his share. At the same time, his wife, who had separated from him three years previously because of his dependence on alcohol, had begun divorce proceedings. The new onslaught of worries had sunk him into a depression which he tried to disguise by consuming up to twelve drinks per day. During the past month before his admission to the hospital he had begun to drink before breakfast while more than doubling his usual intake on weekends, all the while feeling more anxious and suicidal. In the past few weeks he had been unable to recover from his drink-related torments and now his relations with his business partner were almost completely ruptured.

The patient's family history showed another case of alcoholism in a paternal grandfather; one of his father's sisters had died of an overdose of barbiturates. His own pattern of problem drinking had had its origins in college. He had been a solid student in high school, president of the varsity club, and founder of a future businessman's club. When he reached college he was quickly pledged into the most popular fraternity on campus, whose members revelled in drinking contests and held frequent social events at which beer flowed freely. He earned a reputation for his ability to drink great quantities of booze without passing out, and he was proud of it. Before dropping out of college in his third year because of problems stemming from poor attendance, he had experimented with marijuana, hashish, amphetamines, and LSD, but his dependence on alcohol had already been established.

Upon examination at the hospital, Mr Bishop did not give evidence of any other psychiatric problems beside mild, recurrent depressions and a chronic dependence on alcohol. The doctor noted that his fingers and eyelids trembled during the admission interview, he complained of both headache and nausea, and he had much trouble holding his balance and responding to the examiner's questions. He did poorly on simple physical tests like touching his nose with a fingertip. The doctor admitted him for rehabilitation, noting on his record that he appeared to be experiencing early signs of delirium trements (the 'DTs') at the time of admission.

Diagnosing and treating alcoholism is complicated by the fact that it often appears as the stepbrother of other serious illnesses. A person in the early phases of presenile dementia may begin drinking heavily with no forewarning because of a personality change associated with the dementia. It would be quite possible to diagnose the primary disorder as alcoholism, missing the underlying organic brain disease. Alcoholism also frequently accompanies depression, a fact that might help to explain the high rate of association between alcohol abuse and suicide. Drug abusers, notably those who use barbiturates or 'downers', often complicate their problem by developing a dependence on alcohol.

Evidence that alcoholism is inheritable

Family studies of alcoholism all over the world have shown, almost without exception, that alcoholism occurs in relatives of alcoholics at a much higher rate than in the general population. These observations suggested that alcoholism might be genetically inherited. But a familial pattern of disease by itself is not proof of inheritance. Parents and children share similar genes, but in most families, there are also important influences in the environment. Studies were needed to verify the presence of genetic influences and separate them from environmental contributions to alcoholism.

Genetic studies of alcoholism are hindered by several problems. There are so many degrees of severity in alcohol abuse, and so many other mental and physical illnesses can associate with it, that it is hard for a researcher to choose a sample of subjects all having the same disorder. Another barrier to research is that so much alcoholism goes unreported. Some cases of alcoholism are recorded when the drinker creates a public commotion: the drunk driver collides with an oncoming car, an irritable inebriate starts a bar brawl, fighting fuelled by alcohol breaks up a marriage, or medical problems force an alcoholic to seek rehabilitation. But for every case like these, several may go undetected because society indulges the alcoholic, creating a sentimental role for him or her to play, while turning its back on the serious consequences of alcohol abuse. Another barrier to genetic research has been the lack of a research design that could discriminate successfully between environmental and genetic influences in familial transmission of a psychiatric disorder. When twin and adoption studies were applied successfully to genetic research in schizophrenia, a good means was available for studying the role genes might play in alcoholism.

Twin studies of alcoholism

An especially well-designed study of alcoholism was carried out in 1960 by Dr Lennart Kaij, who selected 174 pairs for study from the twin registries of Sweden. Kaij first rated all of the twin subjects on a five-point scale of severity ranging from total abstinence to chronic alcoholism. Then he distinguished several grades of concordance. Twin partners showing identical levels of alcohol use or abuse were rated in the highest grade of concordance. Partners who were one or two levels of severity apart, as when one was dependent on alcohol and the other showed a lesser form of alcohol abuse, were rated in the next lower grade of concordance, and so on down to complete discordance (for example, one partner a non-user, the other addicted to alcohol).

Fraternal-twin pairs in Kaij's study were quite highly concordant for alcohol use; 66.7 per cent of these pairs were rated at the two highest grades of concordance. However, identical-twin pairs showed significantly higher concordance, with 84.5 per cent placed in the two highest grades. The high concordance rate among fraternal twins needs to be explained in part as the result of environmental influences. But the increase in concordance among identical twins has to be explained in part as the result of genetic factors, unless identical-twin pairs are reared in some way that would cause them to be more concordant for alcohol use than fraternal twins (this hypothesis was not tested in the study).

Another important observation made in the Kaij study was that concordance increased among identical-twin pairs along with the severity of alcohol abuse. The greatest concordance among identical twins was in the category of chronic alcoholism. Fraternal-twin pairs, on the other hand, did not show the same kind of pattern. These results suggested to the investigator that severe alcohol abuse and alcohol addicition may be genetically transmitted, while lesser drinking problems do not appear to be.

Twin studies are not designed to distinguish clearly between genetic and environmental influences. Do identical twins drink alike because they are rewarded all their lives for mimicking one another's behaviour? Do identical twins feel a special psychic bond causing them to sympathize with each other's moods? Does being an identical twin impose special stress on a person, inclining him or her to drink? Do the parents of twins differ from other parents in significant ways?

These questions are not easy to answer. And although most researchers do not suspect that such questions hold the key to the results of twin studies, the questions must be asked so that all angles will be covered.

Evidence that alcoholism is inheritable

To separate genetic factors cleanly from the effects of environmental influence, it is necessary to study people who have been separated from their natural parents at birth and raised in adoptive homes.

Adoption studies of alcoholism

The best available adoption studies add considerable weight to the conclusions of twin studies which suggested that there might be genetic components in alcoholism. Furthermore, the adoption studies have shed more light on the relative effects of environmental and hereditary factors in alcoholism, the kind of trait that may be inherited, and the apparent difference in men's and women's susceptibility to alcoholism.

In the early 1970s, a team of Danish and American psychiatrists led by Dr Donald Goodwin, then of St Louis, Missouri, conducted a series of adoption studies of alcoholism in Denmark, using a group of subjects selected from the huge pool of adopted children identified at the start of the landmark Danish adoption studies of schizophrenia (see Chapter 4). Denmark offered many advantages for such a study. The adoption records were well kept and accessible to researchers, as were the registries of psychiatric hospitalizations. And, because the Danish population is not very mobile, members of the sample population were relatively easy to trace.

The Danish adoption studies of alcoholism had three major parts. The first was a study of adopted sons of alcoholics, comparing them against adopted sons whose biological parents were not alcoholic. The second study compared adopted-away sons of alcoholics with their brothers who had not been adopted but were raised by the alcoholic parent. The third study was an investigation of alcoholism and depression in the adopted-away daughters of alcoholics.

Alcoholism in adopted-away sons of alcoholics

In the first study (1973), Goodwin and his fellow workers singled out 55 men (called 'probands' in the study) between the ages of 23 and 45. Each proband had a parent, usually a father, who had once been hospitalized for alcoholism, but each had been placed in the home of biologically unrelated parents before the age of six weeks and afterwards had no further contact with the alcoholic parent. The researchers chose a control group of 50 adopted men who differed significantly from the probands only in that their biological

119

parents were not known to have been hospitalized for alcoholism.

Each of the men underwent a long interview which provided a wealth of information about their ages, marital history, education, level of prosperity, characteristics of foster parents, possible psychiatric problems and psychiatric treatment, and above all, about their patterns of drinking.

Several interesting findings came out of this study. Despite the fact that the probands were raised away from their alcoholic parents, they had more total drink-related problems than the controls. Right down the list—hallucinations, periods of inability to stop drinking, blackouts, tremors, the DTs, rum fits (convulsions), marital and job trouble, arrests, hospitalizations and other treatments for drinking—the proband group, almost entirely without exception, showed a higher rate of problems than the controls. In four of these categories (hallucinations, inability to stop drinking, morning drinking, and treatment for drinking) the increase among the probands was statistically significant.

The researchers also achieved interesting results when they compared overall drinking patterns between the two groups. They defined four categories for comparison: moderate drinker, heavy drinker, problem drinker, and alcoholic. The moderate drinker was defined as neither a teetotaller nor a heavy drinker. A heavy drinker was one who drank more than a certain designated amount over the course of a year (e.g. six more drinks a few times a month) without reporting any problems. A problem drinker was one who qualified as a heavy drinker and who also experienced problems related to alcohol, though not enough to qualify as an alcoholic. An alcoholic was defined as a heavy drinker who experienced problems in three out of four areas, including personal affairs, civic and occupational affairs, physical and mental health, and control of alcohol consumption.

Almost half of the probands and half of the controls were moderate drinkers. Approximately a third in each group were heavy drinkers at some time in their life. About a tenth in each group were problem drinkers at one time. Neither group differed significantly from the other in any of these categories. If anything, in fact, the controls, whose natural parents were not alcoholics, were heavier consumers of alcohol than the probands, although the difference was not statistically significant (i.e. it might be possible to account for the difference by chance).

Evidence that alcoholism is inheritable

But the picture was quite different in the highest category of alcohol abuse, labelled 'alcoholism'. Nearly one out of five of the probands was alcoholic at some time, compared to only one out of twenty of the controls. That means that the probands had nearly four times the rate of alcoholism as the control group. The rate of alcoholism (heavy drinking with severe alcohol-related problems) was the only clear distinction between the drinking patterns of the two groups.

Goodwin and his colleagues concluded from this study that children of alcoholics are more likely to have alcohol problems than children of non-alcoholics, even when both are separated from their alcoholic parents early in life. The study suggested that moderate and even heavy drinking do not appear to be under genetic influence, but that genes may help to transmit the severe drinking pattern compounded with major problems.

The effect of living with an alcoholic parent

The first Danish adoption study of alcoholism supported twin studies in suggesting that genetic transmission of alcoholism may be confined to the most severe forms of the disorder. But these studies could not separate genetic factors in alcoholism from the effects of broken homes, financial troubles, mentally ill relatives, educational disadvantage, legal problems, or whatever other environmental conditions one might suspect of being associated with alcoholism.

The second Danish alcoholism study was designed to tease apart the environmental and hereditary influences. Many of the adopted-away men in the first study had brothers who were not adopted but were raised by the alcoholic parent. This group was of interest because increase in alcoholism among them rather than their adopted-away brothers would have to be attributed to environmental rather than genetic causes. Goodwin and his colleagues chose 20 of the adopted men and 35 of their unadopted brothers for the study.

For many reasons, we might expect the unadopted sons of alcoholic parents to have a higher rate of alcoholism than their adopted-away brothers. First, since the unadopted sons were about three years older on the average, they were further into the standard risk period for alcoholism (roughly ages 20–45). Furthermore, because the unadopted sons grew up in the company of their alcoholic parent or parents, they presumably saw more drinking and were affected by problems related to parental alcoholism more often than their adopted-away brothers. The demographic data on the subjects showed that the unadopted sons belonged to a generally lower socio-economic stratum than their adopted-away brothers. Their alcoholic parents, in turn, belonged to a lower class than the adoptive parents of their brothers. Many studies have noted an association between lower socio-economic status and increased problems with alcohol. Finally, Danish welfare, school, and legal records showed that the unadopted sons had more disruptions in childhood than their adopted-away brothers, including contacts with youth care organizations, problems at school, and truancy. Studies have shown that all of these conditions in childhood are associated with a higher risk of alcoholism in later life.

It comes as something of a surprise, then, that the researchers did *not* find significantly more alcohol problems in the unadopted sons. In fact if any tendency appeared, it was for the adopted-away sons of alcoholics to show more drinking problems and patterns of heavier use than unadopted sons, although the slight differences might have resulted from chance. The researchers concluded that sons of alcoholics were no more likely to become alcoholic if they were reared by their alcoholic parent than if they were separated soon after birth and reared by non-relatives. To put it another way, simply living with an alcoholic parent appeared to have no relationship to the development of alcoholism in the child.

The two Danish studies do not eliminate the possibility that mild alcoholic disorders in men are caused by environmental factors, possibly in combination with biological factors. But they clearly suggest that severe alcoholism in men is strongly influenced by genetic factors.

A genetic study of alcoholism in women

Both of the adoption studies described so far have dealt with sons of alcoholics. Men were singled out because all studies show that the rate of alcoholism is much higher among men than among women. Yet the question of heredity of alcoholic traits among women has interesting angles of its own.

Winokur *et al.* (1975) had suggested that when women developed depression before age 40, their male relatives showed an increase in alcoholism and sociopathy (but not in depression) while their female relatives showed an increase in mood disorders (but not in alcoholism). They coined the term 'depression spectrum disease' for the disorder typified the early onset depression in women. This hypothesis raised the further supposition that among the relatives of alcoholics, women might statistically compensate for their low rate of alcoholism (compared to men) by having a higher rate of mood disorder.

To shed light on such questions, the third Danish adoption study by Goodwin *et al.* (1977) singled out 49 adopted-away daughters of alcoholics and compared them with 47 daughters of non-alcoholics who were also adopted away early in life. Blind, personal interviews were used to collect the data.

As expected, both groups of adopted women had quite low rates of alcoholism in comparison with the adopted men in the earlier studies. Less expectedly, however, the daughters of alcoholics did not have more alcoholism and problem drinking than the daughters of non-alcoholics. Nor did they have a significantly higher rate of depression and other psychiatric illnesses. This study was not designed to prove or disprove the presence of genetic factors in female alcoholism, but it certainly gave no support to the genetic hypothesis.

This study raises many questions. The authors pointed out that they knew little about the parents of control adoptees beside the fact that they were not hospitalized for alcoholism. The possibility remains that the parents were alcoholic and had sought other forms of treatment without hospitalization. This possibility could seriously have affected the results of the study.

Trying to account for the low rate of alcoholism in the daughters of alcoholics in their study, the authors entertained several hypotheses. Perhaps alcoholism in women is not influenced by hereditary factors. Or perhaps a gene for alcoholism is present but the trait is never manifested on the behavioural level because cultural mores in a country like Denmark discourage alcohol use among women. Again, women may differ physiologically from men in response to alcohol, causing them to crave it less or to have less tolerance for it. Finally, the women in the study might have been too young to have developed drinking problems; although the average age of the women was 35 years at the time of the study, possibly 10 or 20 years from now the same subjects would give back quite a different picture.

Alcoholism

X-linkage in alcoholism: pro and con

The authors did not mention the hypothesis that alcoholism in women is a sex-linked trait caused by a recessive gene on the X chromosome so that some of the women in the study may have been carriers of the damaging genes without expressing the trait. The possibility of X-linkage in alcoholism has been debated for a long time among psychiatric geneticists. The suspicion naturally arises from the fact the men consistently have higher rates of alcoholism than women. The fortunes of the X-linked hypothesis have waxed and waned over the years as various studies have produced contradictory evidence. Today the evidence points away from X-linkage, but the question is not yet settled.

The greatest support for the X-linked hypothesis came from a series of four articles by Dr R. Cruz-Coke in the prestigious medical journal, *Lancet*. The first two articles claimed to find an association between colour blindness and cirrhosis of the liver, a frequent complication of chronic alcohol abuse. The second two reported finding an association between colour blindness and alcohol addiction. The gene for colour blindness is known to be located on the X chromosome; when a significant association is found between colour blindness and another trait we have good evidence that the trait is also caused by a gene on the X chromosome. Later studies have explained the association of colour blindness with alcohol addiction as the result of a short-term, probably toxic effect of alcohol on a person's vision. This evidence refutes some of the arguments for X-linkage; yet other studies have turned up new evidence in favour of the hypothesis, thus reviving the debate.

There are numerous ways to test for X-linkage in alcoholism, but one of the most direct is to start with men who are known to have been alcoholics and to study the patterns of alcoholism in their grandchildren, distinguishing between grandchildren produced by daughters and those produced by sons of the alcoholic patriarch. If alcoholism is X-linked (recessive), there ought to be much more alcoholism in the sons of the daughters of alcoholics than in the sons of the sons of alcoholics. The reason for this is quite easy to understand if you remember the functions of the sex chromosomes (the ones labelled X and Y). A man has both an X and a Y in his complete set of genes; he is a male by virtue of that fact. If he passes on a Y chromosome at the time of conception, the offspring will be a son. Since he cannot pass on an X chromosome to a son, no son of his

124

could inherit an X-linked gene for alcoholism from him. Since the son cannot receive one from the father, neither can he pass one on to his son (or daughter). Thus, the grandson of an alcoholic—if he is the son of the alcoholic's son—cannot receive a gene for alcoholism from the father if that gene is X-linked.

But if the alcoholic man passes on an X chromosome at conception, his offspring will be a daughter. That daughter will have two X chromosomes, one from her father and one from her mother. The one from the mother will probably carry a normal gene, since the normal gene would undoubtedly be more common in the population than the allele that would influence the alcoholic trait. The one from the father will, of course, carry a gene for alcoholism. If the daughter inherits an X-linked recessive gene for alcoholism from her father, she will probably be an unaffected carrier. But when she has a son of her own, she will have a 50–50 chance of passing on to him the gene for alcoholism which she inherited from her father. If her son inherits the gene for alcoholism, he will inherit the trait, even if the X-linked gene is recessive, since the small Y chromosome would not contain any corresponding allele to overrule its effects. All things considered, an X-linked gene should cause higher rates of alcoholism in the sons of daughters than in the sons of sons of alcoholics.

This line of reasoning provided the rationale for a study of grandsons of alcoholics by Kaij *et al.* (1975) in Malmo, Sweden. From local registries of problem drinkers, the researchers identified 75 alcoholic men whose sons collectively had borne them 136 grandsons over the age of 15 at the time of the study and whose daughters had produced 134 grandsons in the same age range. The authors compared the two groups of grandsons to see whether there would be significantly more alcoholism in the sons of the daughters. There was not.

The researchers found nearly the same rate of alcoholism in each group of grandsons. The main hypothesis of X-linkage, they concluded, would have to be rejected, at least in the case of their study population. The results of the comparison could not be explained by X-linkage. However, when the researchers examined the rate of alcoholism in the oldest group of grandsons (those who had passed furthest through the standard period of risk for alcoholism), they found that 43 per cent of them had become alcoholic like their grandfathers. Kaij and his colleagues concluded that while such a high rate tends to rule out the possibility of a recessive gene, it might be explainable in terms of an autosomal dominant gene.

The possibility remains that only a certain proportion of cases of alcoholism are transmitted by X-linked genes; others may be transmitted by autosomal dominant or multiple genes, and others may have non-genetic origins. In certain very large families, where the family history of alcoholism and other traits such as blood types or colour blindness is known for several generations, the psychiatric genetic counsellor may be able to discover an unambiguous pattern of X-linkage or autosomal dominance after a careful diagnostic study and completion of a pedigree. But for most cases of alcoholism, the family information will be too incomplete and out current knowledge of genetic components in alcoholism too inadequate to identify a specific genetic mode of transmission. The lack of such information adds one more degree of uncertainty to genetic counselling for alcoholism. But the studies we have discussed in this chapter can offer people attending genetic counselling sessions a broad sense of direction, inasmuch, as they show genetic transmission clearly only for the most severe forms of alcohol abuse and only among males. In lieu of exact risk estimates, the genetic counsellor can supply empirical estimates based on family studies and surveys of the prevalence of alcoholism in the general population.

What is the risk of becoming alcoholic?

The lifetime risk of alcoholism among relatives of alcoholics is unquestionably high, especially among males. Let us set the familial risks of alcoholism in relief against a reasonable estimate of lifelong risk of alcoholism in members of the general population. A survey of studies in Germany, Switzerland, Sweden, Denmark, and England (Goodwin 1976) finds that the lifelong expectancy rate for alcoholism among males appears to be about 3–5 per cent and for females about 0.1–1 per cent. Both of these figures correspond to the estimates for the general population found in a family study in the midwestern United States (Winokur and Tsuang 1978). This study used strict diagnostic criteria and blind, personal interviews in the gathering of data. In both of these sets of data, males in the general population show about a five times higher risk of alcoholism than females.

The rates of alcoholism among male and female relatives of alcoholics are several times higher than the general population risks, but they remain in approximately the same 5:1 proportion between affected males and females.

A pool of six major studies of chronic alcoholics (Rosenthal 1970) shows that the rates for both men and women relatives are more than five times higher than the respective male and female rates in the general population. This risk to fathers of alcoholics ranges from about 11–33 per cent and to brothers of alcoholics from 12–28 per cent. The risk to mothers of alcoholics ranges from 0.4–8 per cent and is slightly higher for sisters of alcoholics, ranging from about 3–10 per cent.

A large family study of alcoholism in St Louis, Missouri, (Winokur *et al.* 1970), using personal interviews and strict diagnostic criteria found still higher rates of alcoholism in close blood relatives of alcoholics. When the alcoholic family members were male, their brothers and sons showed 46 per cent and 31 per cent risk of alcoholism, respectively; their sisters showed a rate of 5 per cent, and no alcoholism was found among their daughters. When the alcoholic family members were male, their brothers and sons showed 46 per cent and 31 per cent risk of alcoholism, respectively; their sisters showed a rate of 5 per cent, and no alcoholism was found among their daughters. When the alcoholic subject was female, 50 per cent of both their brothers and sons were alcoholic, compared to 8 per cent of their sisters and one of their daughters. These higher rates of disorder may be the more reliable ones because of the systematic methods of research used.

As mentioned above, the rate among grandsons of male alcoholics in the Kaij study was 43 per cent.

Lessening the risk and burden of alcoholism

The family studies of alcoholism show that male relatives of alcoholics have a serious risk of inheriting the disorder, a risk possibly approaching the inheritance rate of a classic autosomal dominant trait. And yet, persons who need genetic counselling to help them face the risk of alcoholism seldom come to the counsellor wondering whether or not to bear children. Since alcoholism usually does not set in before early middle age, it affords the affected person a long measure of healthy life. Nor is alcoholism as severely stigmatized in society as other forms of mental disorder.

Many public and private organizations, the most well known of which is Alcoholics Anonymous, offer detoxification programmes that help many alcoholics control or eliminate alcohol use. Some alcoholics are helped by a drug called disulfuram, or Antabuse, which makes drinking of alcohol unpleasant to the addict by causing sweating, fever, and nausea.

Those who need genetic counselling for alcoholism often wish to know who in their immediate family is most at risk for alcoholism. Parents who learn that their children have a high risk of alcoholism may be able to minimize the risk by limiting the access of family members to alcohol and by informing the child at an appropriate time in his life of the special risk of alcoholism he or she has because of the high inheritability of severe alcoholic disorders. Unfortunately, researchers do not yet know how family and social environment contribute to the risk of alcoholism. The Danish study of adopted and non-adopted sons of alcoholics by Goodwin *et al.* (1974) casts some shadow on the assumption that environmental factors play a major role in the transmission of severe alcoholism.

Summary

Alcoholism, too often benignly ignored or sentimentalized in society, is a burdensome disorder affecting as many as one out of 100 women and one out of 20 men in the general population. Family studies find a fivefold increase in the rate of alcoholism among relatives of alcoholics. The familial pattern is caused in part by the transmission of damaging genes. Evidence of genetic influence has come from twin studies, which report 85 per cent concordance for alcoholism among identical twins, and adoption studies of male alcoholics, which have found that removal from the home environment of an alcoholic parent does not eliminate the risk of alcoholism in the male children. However, twin and adoption studies give evidence of genetic influence only for severe alcoholism. One adoption study has found that sons of alcoholics who grow up in the home of the alcoholic parent have no greater risk for alcoholism than their adopted-away brothers. This suggests that environmental factors may not contribute significantly in at least some cases of severe alcoholism. An adoption study of adopted women found no evidence that alcoholism in women is inherited.

The preponderance of males in all surveys of alcoholism suggests the possibility that alcoholism is transmitted by an X-linked recessive gene. A study of grandsons of alcoholics gave no support to this hypothesis, but other studies report an association between colour

Summary

blindness and alcohol addiction, thus supporting an X-linked hypothesis. Some cases of alcoholism may be X-linked while others may have autosomal dominant or polygenic inheritance or non-genetic modes of transmission.

The risks of alcoholism to the relatives of alcoholics are consistently high, with males having five times greater risk of alcoholism than female relatives. The risk to brothers, sons, and grandsons of severe alcoholics appears to be between 40 per cent and 50 per cent. The burden of alcoholism can be lightened in many cases by rehabilitation or medical treatment.

References

American Psychiatric Association (1980). *Diagnostic and statistical manual of mental disorders*, 3rd edn (DSM-III). American Psychiatric Association, Washington, DC.

Angst, J. (1966). Zur ätiologie und nosologie endogener depressiver psychosen. *Monogrn Gesamtgeb. Neurol. Psychiat.* 112.

—— and Perris, C. (1972). The nosology of endogenous depression. *Int. J. ment. Hlth* 1, 145–58.

Böök, J. A. (1953). A genetic and neuropsychiatric investigation of a north-Swedish population. *Acta genet. Statist. med.* 4, 1–100.

Burch, P. R. J. (1964). Manic-depressive psychosis: some new aetiological considerations. *Br. J. Psychiat.* 110, 808–17.

Cadoret, R. J. (1978a). Evidence for genetic inheritance of primary disorder in adoptees. *Am. J. Psychiat.* 135, 463–6.

—— (1978b). Psychopathology in adopted-away offspring of biologic parents with antisocial behavior. *Archs gen. Psychiat.* 35, 176–84.

—— Winokur, G., and Clayton, P. J. (1970). Family history studies: III. Manic depressive disease versus depressive disease. *Br. J. Psychiat.* 116, 625–35.

Cazzullo, C. L., Smeraldi, E., and Penati, G. (1974). The leucocyte antigenic system HL-A as a possible genetic marker of schizophrenia. *Br. J. Psychiat.* 125, 25–7.

Clayton, P. J., Rodin, L., and Winokur, G. (1968). Family history studies: III. Schizo-affective disorder, clinical and genetic factors including a one to two year follow-up. *Compreh. Psychiat.* 9, 31–49.

Cohen, S. M., Allen, M. G., Pollin, W., and Hrubec, Z. (1972). Relationship of schizo-affective psychoses to manic-depressive psychosis and schizophrenia. *Archs gen. Psychiat.* 26, 539–45.

Coppen, A., Montgomery, S. A., Gupta, R. K., and Bailey, J. E. (1976). A double-blind comparison of lithium carbonate and naprotiline in the prophylaxis of the affective disorders. *Br. J. Psychiat.* 128, 479–85.

Crowe, R. R. (1975). Adoption studies in psychiatry. *Biol. Psychiat.* 101, 353–71.

—— (1978). Is genetic counseling appropriate for psychiatric illnesses? In *Controversy in psychiatry* (ed. J. P. Brody, and H. K. E. Brodie). W. B. Saunders, Philadelphia.

Cruz-Coke, R. (1964). Colour-blindness and cirrhosis of the liver. *Lancet* ii, 1064–5.

—— (1965). Colour-blindness and cirrhosis of the liver. Lancet i, 1131–3.

130

References

—— and Varela, A. (1965). Colour-blindness and alcohol addiction. *Lancet* **ii**, 1348.

—— —— (1966). Inheritance of alcoholism: its association with colour-blindness. *Lancet* **ii**, 1282–4.

Curnow, R. N. and Smith, C. (1975). Multifactorial models for familial diseases in man. *Jl R. statist. Soc.* **A138**, 131–56.

Dunner, D. L., Gershon, E., and Goodwin, F. K. (1976). Heritable factors in the severity of affective illness. *Biol. Psychiat.* **11**, 31–42.

Edwards, J. H. (1960). The simulation of mendelism. *Acta genet. Statist. med.* **10**, 63–70.

Elston, R. C. and Campbell, M. A. (1970). Schizophrenia: evidence for the major gene hypothesis. *Behav. Genet.* **1**, 3–10.

—— and Stewart, J. (1971). A general model for the genetic analyses of segregation data. *Hum. Hered.* **21**, 523–42.

—— and Yelverton, K. C. (1975). General models for segregation analyses. *Am. J. hum. Genet.* **27**, 31–45.

—— Kringlen, E., and Namboodiri, K. K. (1973). Possible linkage relationships between certain blood groups and schizophrenia or other psychoses. *Behav. Genet.* **3**, 101–6.

Erlenmeyer-Kimling, L. (1978). Genetic approaches to the study of schizophrenia: the genetic evidence as a tool in research. In *Birth defects original article series*, vol. 14, pp. 59–74. Williams and Wilkins for the National Foundation-March of Dimes, Baltimore.

—— and Paradowski, W. (1966). Selection and schizophrenia. *Am. Nat.* **11**, 651–65.

—— Nicol, S., and Rainer, J. D. (1969). Changes in fertility rates of schizophrenia patients in New York State. *Am. J. Psychiat.* **125**, 916–27.

Falconer, D. S. (1965). The inheritance of liability to certain diseases estimated from the incidence among relatives. *Ann. hum. Genet.* **29**, 51–76.

—— (1967). The inheritance of liability to diseases with variable age of onset, with particular reference to diabetes mellitus. *Ann. hum. Genet.* **31**, 1–20.

Feighner, J. P., Robins, E., Guze, S. B., Woodruff, R. A., Winokur, G., and Munoz, R. (1972). Diagnostic criteria for use in psychiatric research. *Archs gen. Psychiat.* **25**, 57–63.

Fieve, R. and Young, M. (1978). Genetic counseling in manic-depressive disease. In *Controversy in psychiatry* (ed. J. P. Brody and H. K. H. Brodie). W. B. Saunders, Philadelphia.

—— Kumbaraci, T., and Dunner, D. (1976). Lithium prophylaxis of depression in bipolar I, bipolar II, and unipolar patients. *Am. J. Psychiat.* **133**, 925–9.

—— Mendlewicz, J., and Fleiss, J. L. (1973). Manic-depressive illness: linkage with the Xg^a blood group. *Am. J. Psychiat.* **130**, 1355–9.

Fonseca, A. F. da (1959). *Analise heredo-clinica das perturbacoes afectivas.* Faculdade de Medecina, Oporto.

Fowler, R. C., Tsuang, M. T., Cadoret, R. J., Monnelly, E., and McCabe,

References

M. (1974). A clinical and family comparison of paranoid and non-paranoid schizophrenics. *Br. J. Psychiat.* **124**, 346–51.

Fraser, F. C. (1970). Counseling in genetics: its intent and scope. In *Genetic counseling. Birth defects original article series* (ed. D. Bergsma), vol. 1, p. 7. Williams and Wilkins, for the National Foundation-March of Dimes, Baltimore.

—— (1974). Genetic counseling. *Am. J. hum. Genet.* **26**, 636–59.

Garrone, G. (1962). Etude statistique et génétique de la schizophrénie á Genéve de 1901á 1950. *J. Génét. hum.* **11**, 89–219.

Gershon, E. S., Bunney, W. E., Leckman, J. F., Van Eerdewegh, M., and Debauche, B. A. (1976). The inheritance of affective disorders: a review of data and of hypotheses. *Behav. Genet.* **6**, 227–61.

—— Targum, S. D., Kessler, L. R., Mazure, C. M., and Bunney, W. E. (1977). Genetic studies and biologic strategies in the affective disorders. In *Progress in medical genetics* (ed. A. G. Steinberg, A. G. Bearn, A. G. Motulsky, and B. Childs), vol. II. W. B. Saunders, Philadelphia.

—— Mark, A., Cohen, N. Belizon, M., Baron, M., and Knobe, K. E. (1975). Transmitted factors in the morbid risk of affective disorders: a controlled study. *J. psychiat. Res.* **12**, 283–99.

Goetzl, U., Green, R., Whybrow, P., and Jackson, R. (1974). X-linkage revisited. *Archs gen. Psychiat.* **31**, 665–72.

Goodwin, D. W. (1976). Adoption studies of alcoholism. *J. oper. Psychiat.* **7**, 54–63.

—— and Guze, S. B. (1974). Heredity and alcoholism. In *Biology of alcoholism* (ed. B. Kissin and H. Begleiter), vol. 3. Plenum, New York.

—— Shulsinger, F., Hermansen, L., Guze, S. B., and Winokur, G. (1973). Alcohol problems in adoptees raised apart from alcoholic biological parents. *Archs gen. Psychiat.* **28**, 238–43.

—— —— Knop, J., Mednick, S., and Guze, S. B. (1977). Alcoholism and depression in adopted-out daughters of alcoholics. *Archs gen. Psychiat.* **34**, 751–5.

—— —— Moller, N., Hermansen, L., Winokur, G., and Guze, S. B. (1974). Drinking problems in adopted and non-adopted sons of alcoholics. *Archs gen. Psychiat.* **31**, 164–9.

Goodwin, F. K. and Ebert, M. H. (1973). Lithium in mania: clinical trials and controlled studies. In *Lithium: its role in psychiatric research and treatment* (ed. S. Gershon and B. Shopsin). Plenum Press, New York.

Gottesman, I. I. and Shields, J. (1967). A polygenic theory of schizophrenia. *Proc. natn Acad. Sci. U.S.A.* **58**, 199–205.

—— —— (1972). *Schizophrenia and genetics: a twin study vantage point.* Academic Press, New York.

—— —— (1973). Genetic theorizing and schizophrenia. *Bri. J. Psychiat.* **122**, 15–30.

—— —— (1976). A critical review of recent adoption, twin, and family studies of schizophrenia: behavioral genetics perspective. *Schizophren. Bull.* **2**, 360–98.

Green, R., Goetzl, U., Whybrow, P., and Jackson, R. (1973). X-linked

References

transmission of manic-depressive illness. *J. Am. med. Ass.* **223**, 1289.

Hallgren, B. and Sjogren, T. (1959). A clinical and genetico-statistical study of schizophrenia and low-grade mental deficiency in a large Swedish rural population. *Acta psychiat. scand.* Suppl. 140.

Harvald, B. and Hauge, M. (1965). Hereditary factors elucidated by twin studies. In *Genetics and the epidemiology of chronic diseases* (ed. J. V. Neel, M. W. Shaw, and W. U. Schull). PHS Publ. No. 1163, USDHEW, Washington.

Helgason, T. (1964). Epidemiology of mental disorders in Iceland. *Acta psychiat. scand.* Suppl. 173.

Helzer, J. E. and Winokur, G. (1974). A family interview study of male manic depressives. *Archs gen. Psychiat.* **31**, 73–7.

Heston, L. L. (1966). Psychiatric disorders in foster home reared children of schizophrenic mothers. *Bri. J. Psychiat.* **112**, 819–25.

—— (1970). The genetics of schizophrenia and schizoid disease. *Science, N.Y.* **167**, 249–56.

—— and Denney, D. D. (1968). Interaction between early life experience and biological factors in schizophrenia. In *The transmission of schizophrenia* (ed. D. Rosenthal, and S. S. Kety). Pergamon Press, Oxford.

—— and Mastri, A. R. (1977). The genetics of Alzheimer's disease. *Archs gen. Psychiat.* **34**, 976–81.

—— Lowther, D. L. W., and Leventhal, C. M. (1966). Alzheimer's disease: a family study. *Archs Neurol., Chicago* **15**, 225–33.

Huntington, G. (1872). On chorea. *Med. surg. Rep.* **26**, 317–21.

Inouye, E. (1963). Similarity and dissimilarity in twins. *Proc. 3rd Wld Cong. Psychiat.*, vol. 1. University of Toronto Press, Montreal.

James, N. M. and Chapman, C. J. (1975). A genetic study of bipolar affective disorder. *Br. J. Psychiat.* **126**, 449–56.

Jarvik, L. and Chadwick, S. B. (1972). Schizophrenia and survival. In *Psychopathology* (ed. M. Hammer, K. Salzinger, and S. Sutton). Wiley, New York.

Jayakar, S. D. (1970). On the detection and estimation of linkage between a locus influencing a quantitative character and a marker locus. *Biometrics* **26**, 451–64.

Johnson, D. A. W. (1977). Treatment of chronic schizophrenia. *Drugs* **14**, 291–9.

Kaij, L. (1960). *Alcoholism in twins: studies on the etiology and sequels of abuse of alcohol.* Almquist and Wiksell, Stockholm.

—— and Dock, J., (1975). Grandsons of alcoholics: a test of sex-linked transmission of alcohol abuse. *Archs gen. Psychiat.* **32**, 1379–81.

Kallman, F. J. (1950). The genetics of psychoses: an analysis of 1,232 twin index families. Congrés International de Psychiatrie Social, Paris, pp. 1–27.

—— (1952). Human genetics as a science, as a profession, and as a social-minded trend of orientation. *Am. J. hum. Genet.* **4**, 237–45.

—— (1953). *Heredity and mental disorder.* W. W. Norton, New York.

—— (1954). Genetic principles in manic-depressive psychosis. In *Depres-*

References

sion (ed. P. H. Hoch and J. Zubin). Grune and Stratton, New York.

—— (1956). Psychiatric aspects of genetic counseling. *Am. J. hum. Genet.* **8**, 97–101.

Karlsson, J. L. (1973). An Icelandic family study of schizophrenia. *Bri. J. Psychiat.* **123**, 549–54.

Kay, D. W. K. (1978). Assessment of familial risks in the functional psychoses and their application in genetic counselling. *Br. J. Psychiat.* **133**, 385–403.

—— and Roth, M. (1961). Environmental and hereditary factors in the schizophrenias of old age ('late paraphrenia') and their bearing on the general problem of causation in schizophrenia. *J. ment. Sci.* **107**, 649–86.

Kessler, S. (ed.) (1979). *Genetic counseling: psychological dimensions.* Academic Press, New York.

Kety, S. S. (1978). Schizophrenia: the challenge and the prospects of biologic research. In *Birth defects original article series* (ed. D. Bergsma), vol. 14, pp. 5–15. Williams and Wilkins for the National Foundation-March of Dimes, Baltimore.

—— Matthysse, S., and Kidd, K. K. (1978). Genetic counseling for schizophrenic patients and their families. In *Controversy in psychiatry.* (ed. J. P. Brody and H. K. H. Brodie). W. B. Saunders, Philadelphia.

—— Rosenthal, D., Wender, P. H., and Schulsinger, F. (1968). The types and prevalence of mental illness in the biological and adoptive families of adopted schizophrenics. In *The transmission of schizophrenia* (ed. D. Rosenthal and S. S. Kety). Pergamon Press, Oxford.

—— —— —— —— (1971). Mental illness in the biological and adoptive families of adopted schizophrenics. *Am. J. Psychiat.* **128**, 302–6.

—— —— —— —— and Jacobsen, B. (1975). Mental illness in the biological and adoptive families of adopted individuals who have become schizophrenic: a preliminary report based on psychiatric interviews. In *Genetic research in psychiatry* (ed. R. R. Fieve, D. Rosenthal, H. Brill). The Johns Hopkins University Press, Baltimore.

Kidd, K. K. and Cavalli-Sforza, L. L. (1973). An analysis of the genetics of scizophrenia. *Social Biol.* **20**, 254–65.

Kraepelin, E. (1919). *Dementia praecox and paraphrenia.* E. & S. Livingstone, Edinburgh.

Kringlen, E. (1967). *Heredity and environment in the functional psychoses.* Heinemann, London.

Langenbeck, U. and Jorgensen, G. (1976). The genetics of diabetes mellitus—a review of twin studies. In *The genetics of diabetes mellitus* (ed. W. Creutzfeldt, J. Kobberling, and J. V. Neel). Springer-Verlag, Berlin.

Leonhard, K. (1957). *Aufteilung der Endogenen Psychosen*, 1st edn. Berlin.

—— Korff, I., and Schulz, H. (1962). Die temperamente in den Familien der monopolaren und bipolaren phasischen Psychosen. *Psychiatrie neurol. med. Psycol.* **143**, 416.

Loranger, A. W. (1975). X-linkage and manic-depressive illness. *Br. J. Psychiat.* **127**, 482–8.

References

Luxenburger, H. (1942). Das zirkulare Irresein. In *Handbuch der Erbkrankheiten* (ed. A. Gutt). Georg Thieme Verlag, Leipzig.

Lynch, H. T., Guirgis, H., and Bergsma, D. (eds) (1977). *International directory of genetic services*, 5th edn. The National Foundation-March of Dimes, White Plains,

McCabe, M. S., Fowler, R. C., Cadoret, R. J., and Winokur, G. (1971). Familial differences in schizophrenia with good and poor prognosis. *Psychol. Med.* **1**, 326–32.

McKusick, V. A. (1975). *Mendelian inheritance in man: catalogs of autosomal dominant, autosomal recessive, and X-linked phenotypes*, 4th edn. The Johns Hopkins University Press, Baltimore.

Meehl, P. E. (1962). Schizotaxia, schizotypy, schizophrenia. *Am. Psychol.* **17**, 827–38.

Mendlewicz, J. (1976). The contribution of genetics to biological psychiatry. *Neuropsychobiology* **2**, 65–73.

—— and Rainer, J. D. (1974). Morbidity risk and genetic transmission in manic-depressive illness. *Am. J. hum. Genet.* **26**, 692–701.

—— —— (1977). Adoption study supporting genetic transmission in manic-depressive illness. *Nature, Lond.* **268**, 327–9.

—— Fleiss, J., and Fieve, R. (1972). Evidence for X-linkage in the transmission of manic-depressive illness. *J. Am. Med. Ass.* **222**, 1624–7.

Meyer, N. (1977). Diagnostic distribution of admissions to inpatient services of state and county mental hospitals, United States, 1975. Mental Health Statistical Note No. 138. Alcohol, Drug Abuse and Mental Health Administration, PHS, USDHEW, Rockville, Maryland.

Milunsky, A. (1977). *Know your genes*. Houghton Mifflin, Boston.

Mitsuda, H. (1972). Heterogeneity of schizophrenia. In *Genetic factors in 'schizophrenia'*. (ed. A. R. Kaplan). Charles C. Thomas, Springfield, Ill.

Murphy, E. A. (1973). Probabilities in genetic counseling. In *Contemporary genetic counseling. Birth defects original article series* (ed. D. Bergsma), vol. 9, pp. 19–33. Williams and Wilkins for the National Foundation-March of Dimes, Baltimore.

—— and Chase, G. A. (1975). *Principles of genetic counseling*. Year Book Medical Publishers, Chicago.

National Institute of Mental Health (1977). Additions and resident patients at end of year, state and county mental hospitals by age and diagnosis, by state, United States, 1975. National Institute of Mental Health, Rockville, Maryland.

Neel, J. V. (1976). Diabetes mellitus—a geneticist's nightmare. In *The genetics of diabetes mellitus*. (ed. W. Creutzfeldt, J. Kobberling, and J. V. Neel). Springer-Verlag, Berlin.

Nora, J. J. and Fraser, F. C. (1974). *Medical genetics: principles and practice*. Lea and Febiger, Philadelphia.

Odegaard, O. (1952). La genetique dans la psychiatrie. *Proc. 1st Wld Cong. Psychiat.* Hermann, Paris.

—— (1972). The multifactorial theory of inheritance in predisposition to schizophrenia. In *Genetic factors in 'schizophrenia'* (ed. A. R. Kaplan).

References

Charles C. Thomas, Springfield, Ill.

Peet, M. (1975). Lithium in the acute treatment of mania. In *Lithium research and therapy* (ed. F. Johnson). Academic Press, London.

Penrose, L. S. (1956). Some notes on heredity counselling. *Acta genet. Statist. med.* **6**, 35–40.

Perris, C. (1966). A study of bipolar (manic-depressive) and unipolar (recurrent depressive) psychoses. *Acta psychiat. scand.* Suppl. 194.

—— (1968). Genetic transmission of depressive psychoses. *Acta psychiat. scand.* Suppl. 203.

—— (1971). Abnormality on paternal and maternal sides: observations in bipolar (manic-depressive) and unipolar depressive psychoses. *Br. J. Psychiat.* **118**, 207–10.

—— (1974). The genetics of affective disorder. In *Biological psychiatry* (ed. J. Mendels). Wiley, New York.

Price, J. (1968). The genetics of depressive behavior. In *Recent developments in affective disorders* (ed. A. Coppen and A. Walk), *Br. J. Psychiat.*, Special Publ. **2**, 37–54.

Prien, R. F., Caffey, E. M., Jr, and Klett, C. J. (1973). Lithium carbonate and imipramine in prevention of affective episodes. *Archs gen. Psychiat.* **29**, 420–5.

Rainer, J. D. (1966). Genetic counselling and preventive psychiatry. *Ment. Hyg. Lond.* **50**, 593–5.

Reich, T., Cloninger, C. R., and Guze, S. B. (1975). The multifactorial model of disease transmission: I. Description of the model and its use in psychiatry. *Bri. J. Psychiat.* **127**, 1–10.

—— James, J. W., and Morris, C. A. (1972). The use of multiple thresholds in determining the mode of transmission of semi-continuous traits. *Ann. hum. Genet.* **36**, 163–84.

Robertson, A. (1973). Linkage between marker loci and those affecting a quantitative trait. *Behav. Genet.* **3**, 389–91.

Rosanoff, A. J., Handy, L. M., and Plesset, I. R. (1935). The etiology of manic-depressive syndromes with special reference to their occurrence in twins. *Am. J. Psychiat.* **91**, 725–62.

Rosenthal, D. (1970). *Genetic theory and abnormal behavior.* McGraw-Hill, New York.

—— (1971). A program of research on heredity in schizophrenia. *Behav. Sci.* **16**, 191–201.

—— (1972). Three adoption studies of heredity in the schizophrenic disorders. *Int. J. ment. Hlth* **1**, 63–75.

—— Wender, P. H., Kety, S. S., Schulsinger, F., Welner, J., and Ostergaard, L. (1968). Schizophrenics, offspring reared in adoptive homes. In *The transmission of schizophrenia* (ed. D. Rosenthal and S. S. Kety). Pergamon Press, Oxford.

Schenk, V. W. D. (1959). Re-examination of a family with Pick's disease. *Ann. hum. Genet.* **23**, 325–33.

Schou, M. (1973). Prophylactic lithium maintenance treatment in recurrent endogenous affective disorders. In *Lithium: its role in psychiatric research*

References

and treatment (ed. S. Gershon, and B. Shopsin). Plenum Press, New York.

—— and Thomsen, K. (1975). Lithium in the prophylactic treatment of recurrent affective disorders. In *Lithium research and therapy* (ed. F. N. Johnson). Academic Press, London.

Schultz, B. (1932). Zur erbpathologie der Schizophrenia. *Z. ges. Neurol. Psychiat.* **143**, 175–293.

Shields, J. (1962). *Monozygotic twins brought up apart and brought up together.* Oxford University Press, London.

—— (1976). Genetics in schizophrenia. In *Schizophrenia today* (ed. D. Kemali, G. Bartholini, and D. Richert). Pergamon Press, Oxford.

—— and Slater, E. (1967). Genetic aspects of schizophrenia. *Hosp. Med.* April, 579–84.

Shopsin, B., Mendlewicz, J., Suslak, L., Silbey, E., and Gershon, S. (1976). Genetics of affective disorders: II. Morbidity risk and genetic transmission. *Neuropsychobiology* **2**, 28–36.

Simpson, N. E. (1976). The genetics of diabetes mellitus—a review of family data. In *The genetics of diabetes mellitus.* (ed. W. Creutzfeldt, J. Kobberling, and J. V. Neel). Springer-Verlag, Berlin.

Sjogren, T., Sjogren, H., and Lindgren, A. G. H. (1952). Morbus Alzheimer and morbus Pick: a genetic, clinical and patho-anatomical study. *Acta psychiat. scand.* Suppl. 82.

Slater, E. (1938). Zur Erbpathologie des manisch-depressiven Irreseins: die Eltern und Kinder von Manisch-Depressiven. *Z. ges. Neurol. Psychiat.* **163**, 1–147.

—— (1953). Psychotic and neurotic illness in twins. Medical Research Council Special Report Series, No. 278. HMSO, London.

—— (1958). The monogenic theory of schizophrenia. *Acta genet. Statist. med.* **8**, 50–6.

—— and Cowie, V. (1971). *The genetics of mental disorder.* Oxford University Press, London.

—— and Tsuang, M. T. (1968). Abnormality on paternal and maternal sides: Observations in schizophrenia and manic-depression. *J. med. Genet.* **5**, 197–9.

—— Maxwell, J., and Price., J. S. (1971). Distribution of ancestral secondary cases in bipolar affective disorders. *Br. J. Psychiat.* **118**, 215–18.

Sly, W. S. (1973). What is genetic counseling? In *Contemporary genetic counseling. Birth defects original article series* (ed. D. Bergsma), vol. 9, pp. 5–18. Williams and Wilkins for the National Foundation-March of Dimes, Baltimore.

Smith, C. (1971). Recurrence risks for multifactorial inheritance. *Am. J. hum. Genet.* **23**, 578–88

Sperber, M. A. and Jarvik, L. F. (eds.) (1976). *Psychiatry and genetics: social, ethical, and legal implications.* Basic Books, New York.

Stenstadt, A. (1952). A study in manic-depressive psychoses: clinical, social and genetic investigations. *Acta psychiat. scand.* Suppl. 79.

References

Stephens, J. H. and Astrup, C. (1963). Prognosis in 'process' and 'Non-process' schizophrenia. *Am. J. Psychiat.* **119**, 945–51.

Stern, C. (1973). *Principles of human genetics.* W. H. Freeman, San Francisco.

Stevenson, A. C., Davison, B. C. C., and Oakes, M. W. (1970). *Genetic counseling.* Lippincott, Philadelphia.

Taylor, M. and Abrams, R. (1973). Manic states. *Archs gen. Psychiat.* **28**, 656–8.

—— —— (1978). The prevalence of schizophrenia: a reassessment using modern diagnostic criteria. *Am. J. Psychiat.* **135**, 945–8.

Trzebiatowska-Trzeciak, O. (1977). Genetical analysis of unipolar and bipolar endogeneous affective psychoses. *Br. J. Psychiat.* **131**, 478–85.

Tsuang, M. T. (1965). *A study of pairs of sibs both hospitalized for mental disorder,* Ph.D. Thesis, University of London.

—— (1967). A study of pairs of sibs both hospitalized for mental disorder. *Br. J. Psychiat.* **113**, 283–300.

—— (1971). Abnormality on paternal and maternal sides in Chinese schizophrenics. *Br. J. Psychiat.* **118**, 211–14.

—— (1975a). Genetics of affective disorder. In *The psychobiology of depression* (ed. J. Mendels). Spectrum Publications, New York.

—— (1975b) Heterogeneity of schizophrenia. *Biol. Psychiat.* **10**, 465–74.

—— (1976). Genetic factors in schizophrenia. In *Biological foundations of psychiatry* (ed. R. G. Grennel and S. Gabay). Raven Press, New York.

—— (1977). Genetic factors in suicide. *Dis. nerv. Syst.* **38**, 498–501.

—— (1978). Genetic counseling for psychiatric patients and their families. *Am. J. Psychiat.* **135**, 1465–75.

—— (1979). Schizoaffective disorder: dead or alive? *Archs gen. Psychiat.* **36**, 633–4.

—— and Winokur, G. (1974). Bipolar primary affective disorder. *J. oper. Psychiat.* **6**, 47–53.

—— —— (1974). Criteria for subtyping schizophrenia. *Archs gen. Psychiat.* **31**, 43–7.

—— Dempsey, M. D., and Rauscher, F. (1976). A study of 'atypical schizophrenia'. *Archs gen. Psychiat.* **33**, 1157–60.

—— Leaverton, P. E., and Huang, K. S. (1974). Criteria for subtyping poor prognosis schizophrenia: a numerical model for differentiating paranoid from nonparanoid schizophrenics. *J. psychiat. Res.* **10**, 89.

United States National Center for Health Statistics (1978). *International classification of diseases* (WHO), *9th Revision, Clinical Modification* (*ICD-9-CM*).

Von-Grieff, H., McHugh, P. R., and Stokes, P. (1975). The familial history in sixteen males with bipolar manic depressive disorder. *Proc. Am. Psychopath. Ass.* **63**, 233–9.

Weissman, M. M. and Myers, J. K. (1978). Affective disorders in a U.S. urban community. *Archs gen. Psychiat.* **35**, 1304–11.

—— —— (1978). Rates and risks of depressive symptoms in a United States urban community. *Acta psychiat. scand.* **57**, 219–31.

References

—— —— and Harding, P. S. (1978). Psychiatric disorders in a U.S. urban community, 1975–1976. *Am. J. Psychiat.* **135**, 459–62.

Wender, P. H., Rosenthal, D., and Kety, S. S. (1968). A psychiatric assessment of the adoptive parents of schizophrenia. In *The transmission of schizophrenia* (ed. D. Rosenthal and S. S. Kety). Pergamon Press, Oxford.

—— —— —— Schulsinger, F., and Welner, J. (1974). Crossfostering: a research strategy for clarifying the role of genetic and experimental factors in the etiology of schizophrenia. *Archs gen. Psychiat.* **30**, 112–18.

Winokur, G. (1970). Genetic findings and methodological considerations in manic depressive disease. *Br. J. Psychiat.* **117**, 267–74.

—— and Clayton, P. (1967). Family history studies: two types of affective disorders separated according to genetic and clinical factors. In *Recent advances in biological psychiatry* (ed. J. Wortis), vol. 9. Plenum Press, New York.

—— and Tanna, V. L. (1969). Possible role of X-linked dominant factor in manic-depressive disease. *Dis. nerv. Syst.* **30**, 89–95.

—— and Tsuang, M. T. (1978). Expectancy of alcoholism in a midwestern population. *J. Stud. Alcohol* **39**, 1964–7.

—— Cadoret, R. J., and Dorzab, J. (1971). Depressive disease: a genetic study. *Archs gen. Psychiat.* **24**, 135–44.

—— Clayton, P. J., and Reich, T. (1969). *Manic depressive illness.* (C. V. Mosby, St Louis).

—— Morrison, J., Clancy, J., and Crowe, R. R. (1974). Iowa 500: the clinical and genetic distinction of hebephrenic and paranoid schizophrenia. *J. nerv. ment. Dis.* **159**, 12–19.

—— Reich, T., Rimmer, J., and Pitts, F. N. (1970). Alcoholism: III. Diagnosis and familial psychiatric illness in 259 alcoholic probands. *Archs gen. Psychiat.* **23**, 104–11.

Zerbin-Rüdin, E. (1967). Endogene psychosen. In *Humangenetik, ein Kurzes Handbuch* (ed. P. Becker), vol. 2. Thieme, Stuttgart.

Index

Index

Index